Leadership Throughout
How to create successful enterprise
by Richard Jeffery MA MNI

Geese fly in formation. The leader creates a slip stream and the followers are able to keep up. When the leader tires another takes the lead and the tired goose then travels in the slip stream. When one goose is in difficulty and becomes unable to keep up another will drop out of formation and support her until she is able to rejoin the group. This is leadership throughout the group.

Leadership Throughout

Published by The Nautical Institute
202 Lambeth Road, London SE1 7LQ, England
Telephone +44 (0)207 928 1351 Fax +44 (0)207 401 2817
website: www.nautinst.org
First edition published 2007
Copyright © The Nautical Institute 2007
© Richard Jeffery 2007

All rights reserved. No part of this publication may be reproduced, stored in a retrieval system, or transmitted in any form, by any means, electronic, mechanical, photocopying, recording or otherwise, without the prior written permission of the publisher, except for the quotation of brief passages in review. Although great care has been taken with the writing of the book and production of the volume, neither The Nautical Institute nor the author can accept any responsibility for errors or omissions or their consequences.

The book has been prepared to address the subject of leadership. This should not, however, be taken to mean that this document deals comprehensively with all of the concerns that will need to be addressed or even, where a particular matter is addressed, that this document sets out the only definitive view for all situations. The opinions expressed are those of the author only and are not necessarily to be taken as the policies or view of any organisation with which he has any connection.

Typeset by J A Judd
Tradeset Ltd, 2 Burwood Avenue, Eastcote, Pinner, HA5 2RZ

Printed in England by
Modern Colour Solutions, Hayes Road, Middlesex, UB2 5NB

ISBN 1 870077 85 7

Acknowledgements

I am indebted to many who have made this book turn from a dream to a reality.

My thanks to Julian Parker and the Nautical Institute for commissioning me and for Julian's willingness to offer meaningful support and guidance during the process. I also want to thank the University of Exeter and in particular Keith Kinsella and Jonathan Gosling. Keith inspired me to become rather more reflective and less judgemental of others: a life skill that I know I will not lose. And Jonathon for his support and willingness to write the forward.

The major parts of this book are grounded in the work of several acknowledged experts. I am particularly indebted to the work of Peter Senge, Abraham Maslow and Douglas McGregor who, among many others, have helped clarify my own thoughts and added powerful insights into our attitudes to work and each other.

Finally my gratitude goes to Anne, Catherine and Claire who inspire and support me every day. I am very proud of you.

Leadership Throughout is the third volume in the 'Maritime Futures' series.

Foreword

by Professor Jonathan Gosling

Admiral Lord Horatio Nelson, perhaps the greatest – certainly the best-known – maritime leader, never read a leadership manual. But he read plenty: French naval manuals, technical bulletins on navigation, the accounts of explorers and traders. These all contributed to his continuing excellence at his core trade: that of a seaman and commander. Later in life, as he sailed south to take command of the Mediterranean fleet in 1803, he wrote a far-sighted assessment of the political and military situation, a true strategic overview, illustrating his grasp of current affairs, the imperative of trade, and the nuances of international politics. Crucially, he showed himself able to manage the force at his disposal for long-term strategic effects well beyond the immediate challenges of containing the French fleet. But if that is all we notice- that Nelson had become a strategist – we would miss some of the most important and interesting lessons of his example. I think there are three key things to note.

Firstly, Nelson spent an average of 6 hours per day on correspondence. He gave close attention to gathering and interpreting intelligence, maintaining relationships with informers and friendly governments or factions ashore; he took a personal interest in the promotion and discipline of officers across his fleet; he arranged interesting and lucrative 'cruises' to relieve the stresses of those on blockade duty off Toulon; he wrote an endless stream of letters to the Admiralty and the supply clerks to keep his ships stocked and armed. In other words, he remained an active manager of all the affairs of his fleet and this part of the campaign that led up to Trafalgar was as remarkable as the battle eventually turned out to be: to keep the fleet at sea and effective for two years without a single friendly port to retire to, under sail the entire time and relying on personal contacts for supplies and intelligence. Leadership depends on administration for its effects.

Secondly, amongst his vast correspondence, Nelson wrote countless personal letters to his friends, and of course to his lover, Emma Hamilton. He wrote about his ships and his colleagues, but most of all, he wrote about his own state of mind. He was a remarkably reflective man, acutely aware of his own energy levels, the effect he was having on those around him, the demands he was making of his physical and mental energies. Leadership – especially if it is to be sustained under pressure for long periods – depends on self-awareness.

Thirdly, he stayed clear about his purpose. In all his communications, from informal dinners with his captains through to bureaucratic correspondence with supply clerks he emphasised the big-picture reasons for being there, and at the same time showed his appreciation for the skill and dedication given to each little job: everyone could see how their part fitted into the whole. Discipline was tough, sometimes brutal; but no-one could doubt the work to be done and the need for each to play a part. Leadership derives its meaning from a common task – and depends on constant reaffirmation of that task.

These three lessons – the need for effective management, self-awareness and worthwhile work – are at the heart of this book. These are not simply arbitrary points dreamt up by the author: rather, they are lessons from history, and they will prove useful if they connect with your own experience. There are many other lessons to be learned from this book, as from history; and the further we go in applying and thinking about these lessons, the more intriguing questions we find. That is why one of the most important traits of leadership – and one which Nelson exemplified more than any other – is this: to be fascinated, curious, critical and questioning about it, in oneself and in others. If this book helps you to lead, it will have done half its job; it will do the other half if it makes you intrigued to know more.

Jonathan Gosling, 2007
Director of the Centre for Leadership Studies and Head of Executive Education, School of Business and Economics, University of Exeter.

Preface

Leadership Throughout and the Value of Leadership

Julian Parker

Leadership is one of those terms like development that is dependent for its meaning on circumstances. Leadership has no opposite. We do not have a word called unleadership; instead we have endowed the concept with only positive value. This can be confusing if a leader takes us over a cliff or more probably chooses the wrong market for the company's product; so how can this be resolved?

How many leaders do we need? This is another awkward question in the context of a book devoted to the subject, but if everybody is aspiring to take the initiative, push above their weight and generally set their direction for the organization, there may be more confusion than before.

In spite of the independent status it confers on an individual, a leader can only lead to where sustainable development is possible. Napoleon for example paid for his empire by conquering and appropriating foreign lands. Parliament on the other hand paid Nelson to beat Napoleon's navy through taxation. When the smoke had settled, after the battle of Trafalgar, the outcome created a new order that led in turn to new opportunities which other leaders were able to exploit.

History passes its own judgement, but one can see the natural desire to make the most of opportunity as one of the key driving forces in the human soul. Should we be surprised if this quality has a generic component? I do not think so.

To work to our full potential and enjoy the fruits of our labour would therefore appear to be an inalienable right of all individuals and yet we know that organisations become bogged down with bureaucracy, political in fighting and obsolete practices. We lament "How could it come to this?" but nobody hears our prayers and we wonder what went wrong.

So we start to realize that other forces – dark forces are marauding in the corridors of power, in common rooms and behind the scenes. Jealousy, greed, fear and envy release powerful pheromones to mark their murky territory in which hide the damned guardians of the three ions, protection, extortion and corruption. Could these sinister characteristics have been handed down from our ancestors too? It is hard to know where else they could have come from.

But being protective is very much part of survival. Groups defend themselves against predators, society enacts commandments which are upheld by the rule of law and there are strong emotional responses whenever people are threatened. So where does this leave our leader, out in front or corralling the caravan against hostile forces?

Sometimes those in authority have to be expansive and at other times they have to be defensive, but the quality, which enables him or her to decide which action is most appropriate, is the same. It is the ability to see ahead and evaluate the consequences, but that is not the end of the matter. A prescient person may divine the future but may be too weak to respond, a fantasist may see so many possibilities that he never makes up his mind and the visionary may remain so powerfully focused on just one task that she becomes inflexible. Somewhere in between we have to find our true leader.

It will be evident by now that a true leader has to have courage and backbone. The former is rooted in belief in oneself and the purpose of the venture. The latter provides stability and a perspective from which to assess consequences, but if leaders are too pliable then their role becomes superfluous and nobody takes any notice.

Almost by definition a leader is expected to lead, but is that something we can all do? Of course it is. In our own way and in our own time we become leaders in different situations. The mother who heroically steers her family from poverty, the Rotarian organizing a charity event who is a car mechanic, the teacher who takes the children on a field trip or the cadet who brings his lifeboat full of survivors to a safe haven all exhibit what it takes.

What is more difficult is to see the way forward in an organizational context. So much of education and training is directed to proficiency and the ability to perform well in a chosen discipline, but when the time comes for a worker to be promoted to section head or a practitioner to a manager, their background does not necessarily prepare them for the role ahead. They may well be afraid of such a commitment and seek to avoid the added responsibility. It is difficult for them to imagine how to behave.

Quite apart from the transition within an organisation the brain has to start solving different kinds of problems. The research by Edward de Bono in the 1960's discovered that the brain is a self-organising system that routinely interprets inputs into patterns and that, for most of us; it is not inherently designed to be creative. If we expect leaders to see the position today and envisage a different scenario tomorrow, how are they going to do it without becoming creative?

Fortunately a number of researchers, practitioners and academics have provided some meaningful clues to help us find an answer. They suggest that the brain can be trained so that we can all be creative, if we try and practice hard enough. Whereas it may be hard for some of us to be creative, it need not be impossible.

It is human to be trapped in the manner of our upbringing and in the behaviour we have learnt. It is empowering, liberating, stimulating and joyful to realize that the trap is an illusion and that we are not destined to be confined within the limits of our current apprehensions if we choose not to be.

So, back to our original question. How many leaders do we need? In any group the answer is one. However, this is relative. The company has a CEO, he is the leader whereas the crew has a Chief Petty Officer, he is also the leader. The leader appears at all levels of the organization. As in the case of the CPO most leaders are also followers and in this statement lays the true essence of leadership. Leadership is a collaborative venture, the CEO leads the company through his vision, his example and his senior managers. The senior managers lead their staff while at the same time following the leadership of the CEO and so on throughout the organisation. Leadership alignment thus ensures that leaders at all levels use their individual leadership responsibility in the best interest of the collective group, avoiding the mayhem that occurs when alignment is absent.

However, just appointing a leader is not really a total solution to a management problem. The real effectiveness comes from the followers who support the purpose and contribute to the aim. Without them there is no team to coordinate.

This book is rightly focused on leadership throughout and not at leaders at the top. All members of a group have an obligation to support their leaders. However, leaders at all levels have a similar obligation to inspire and develop their followers. This shared responsibility can be uplifting for leaders and followers alike, one is no more important than the other. Think of an orchestra and its conductor. Are musicians any less for not being on the podium? A good concert is shared by all players and the audience. The combined output of energy and emotion is higher than might be expected. True synergy has been achieved, not because of the conductor but because the leaders and followers were aligned and complementary in their efforts.

Leadership Throughout can be shared by all and when it is, the results will be better than expected too.

Julian Parker OBE BSc FNI FRSA, Former Secretary of The Nautical Institute and Director of Publishing.

LEADERSHIP THROUGHOUT

How to create successful enterprise

Contents

Acknowledgement iii

Forward by Professor Jonathan Gosling iv

Leadership Throughout and the value of leadership by Julian Parker vi

Table of contents ix

Introduction..1

Part one:
Why leadership?

Chapter 1 Setting the leadership scene..7

Summary; What is leadership?; Committing to sustainable leadership; The role of training; The leadership pyramid; The Hand of Leadership

Chapter 2 The Environment and Ethics; leadership benchmarks..........19

Summary; Man's inhumanity to man; The golden rule of ethics

Chapter 3 Leadership versus management..23

Summary; Defining leadership; How leadership and management coexist; The command and control example; Why do we need leadership?; Leadership harnesses the people 27

Chapter 4 The born leader myth: False or true?...29

Summary; The perfect leader; do all leaders have loud voices?; Exposing the myth; Good or bad leadership?; The turnaround specialist

Chapter 5 Becoming a leader..35

Summary; Leadership needs in merchant fleets; The need for more leaders; Liberating the Quiet Voice; McGregor's X and Y model; Dispensing with old habits; Accept the need to change

Chapter 6 Creating leaders: A journey of discovery...................................43

Summary; Confidence and habits; Starting to change; Adjusting the mask; Allowable weaknesses; Personal strengths and weaknesses; Developing awareness and empathy; Leadership reality; The ladder of inference; Organisations and leadership

Chapter 7 Leadership styles 57

Summary; Who makes the best leader?; The trait model; Contingency based leadership theories; Situational leadership; Distributed leadership; The quiet leadership model; Autocratic leadership models; Transactional leadership; Transformational leadership; Skills and behavioural models; A Continuum or a circle of choice?; Changes in leadership thinking; Practising leadership

Part two:

The hand of leadership has five fingers

Chapter 8 Awareness: Of self and others 73

Summary; Behaviour types; Maslow's hierarchy; Behaviour change

Chapter 9 Direction: Goals and values 81

Summary; Giving direction with guidance; Work without goals; Corporate and departmental visions; Setting objectives; Aligning personal and business objectives; Values; Travelling the path; Communicating the vision; Why have values?; Values and time management

Chapter 10 Openness: Two way communications 93

Summary; Listening as a skill and a necessity; Active listening; Questioning; Leadership as a two way dialogue; Communicating objectives; Constant communications; Being consistent; Trust and mutual respect; Integrity; Transparency and clarity; Daring Dialogues; Giving feedback;

Chapter 11 Atmosphere: Avoid de-motivating 115

Summary; Motivation as a quick fix?; Enthusiasm and passion; Optimism; Filling leadership voids; De-layering and bureaucracy

Chapter 12 Action: Developing a leadership culture 125

Summary; Survival versus excellence; Sustaining leadership; Coaching; Sharing; Handling change; Demanding and recognising success; Challenging others; Letting others take the credit; Walking the talk; The freedom to lead; Tough action; Power of leadership;

Conclusion: Embracing change 135

Summary; Embracing change; is there risk in the future?

Bibliography 139

Appendix 1 Communicating with clarity 140

Index 141

Introduction

Leadership covers a vast ocean of possibilities and a single volume can not do the whole subject justice. This book focuses on leadership for the practitioner and avoids strategic leadership which is the domain of Boards, the CEO and the most senior corporate management. In doing so it aims to provide practical advice for practical people.

It is divided into two parts. Chapters 1 to 7 ask "Why Leadership?" This puts the case for becoming a leader and encouraging others to lead before moving on to describe some leadership styles that have been identified over the years. The chapters of the second part turn the discussion turns towards practical leadership and "The Hand of Leadership." Each finger represents a key area of managerial responsibility but it goes on to explain that collectively they are leadership.

Mariners are adventurers by nature and adventurers pursue their dreams with a vigour that others often lack. In this we are particularly fortunate. Sir Ernest Shackleton is the epitome of the mariner adventurer. Having gained his master's certificate by the age of twenty four he went on to become one the great explorers of the 20th century. In ensuring the survival of all his crew and expedition colleagues for one and a half years after their ship, *Endurance*, became stuck and eventually broke up in the Antarctic ice he established his leadership credentials beyond question.

"I do not think there is any doubt that we all owe our lives to his leadership and his power of making a loyal and coherent party out of diverse elements" Reginald W. James, physicist, Endurance. (Morrell 2001)

The myth that leaders cannot be made and that leaders are born lingers in the minds of many and yet the fact is that with effort, guidance and practice we can all develop improved and often exceptional leadership capability. What matters most to aspiring leaders is a passionate urge to keep learning, a determination to succeed, and a will to understand themselves and help others develop. Leadership is common sense.

This book considers leadership wherever it can be effectively practiced in organisations. It proposes that leadership can be used as a tool by everyone from managing directors to fleet managers and from department managers to petty officer and cadets.

The focus is on the leadership skills that are a significant component of the full skill set that is necessary for individuals to improve their leadership impact. Many talk of "soft skills" whilst here they are referred to as "leadership skills." Unfortunately the term "soft skills" is sometimes interpreted as something that is not important, something that can be neglected or ignored, something on the periphery of the business world. The word "soft" invokes feelings of easy and gentle and these adjectives are not reflective of effective leadership. Leadership skills provide the high-octane fuel that is needed to drive any business to sustainable success. Without such fuel the engine may work but it will never be a race winner. They are skills that can and must be learned, practiced and applied. Of course there are born leaders and born managers but that does not mean to say you cannot work on becoming one of the best even without being given it as a birth right.

Popular opinion may suggest that in order to be a good manager, with the ability to lead, you must be a self opinionated extrovert and a skilled politician, in fact a born leader. My first contention is that this is how the self-opinionated political extroverts would like to see it. They typify the arrogance and egocentricity that is associated with some examples of leadership. I can recall, as a young cadet, sailing with one such captain, he commanded his ship with absolute authority. Officers and crew that sailed with him did not try to excel, instead they were obliged to spend their time trying to avoid upsetting the narcissistic ship's master. Such people are often the dominating noise in society and they are too often the only voice that is heard. More than two thousand five hundred years ago Plato struggled with the realisation that society was a free for all where the loudest voices prevailed. The problem he saw was that it was not the cleverest or clearest voice, it was instead usually the one that was heard above the others, that got its way (Grint, 1997). Things do not appear to have changed much over the centuries.

Why on earth do such people hold down responsible positions? Because we are perhaps taken in by those who are able to devour us with rhetoric and first impressions; we do not manage to see beyond the bluster. As an old phenomenon it is hardly surprising that Hannah Arendt a famous 20th century political philosopher once wrote;

"By avoiding the bad and the ugly, we deny our own experience. We negate the fact that bad leadership rather than good dominates the daily news and we ignore the realities of the workplace, where most of us regularly witness management and leadership that is conspicuously less than perfect"

Of those who can see through this façade, many remain silent and inaction has provided the world with a large number of management leaders who are in reality without adequate substance or worse still, incompetent.

Look around you, look in your office or on your recent ships and consider your own experiences of different types of management and leadership. No doubt you have seen some that bear many similarities to the type outlined above. Given this situation it is easy to offer another contention; that there is plenty of room for real leaders to be developed within organisations.

Whether you talk of 'management', 'leadership' or both there is no escaping the fact that the issues surrounding them are common sense applied in a logical and consistent fashion. Leadership is both a management and a situational skill, different styles are needed in different situations.

The effective application of leadership skills is an important key to future success both for individuals and the companies they work with. As managers, officers or supervisors you have come up through some kind of formal training, perhaps in nautical studies, engineering or as a scientist or accountant and this training equips you to start your career. Over the years you have seen yourself evolve into functions requiring more managerial and by default, leadership skills. This evolutionary process is sometimes coupled with a coordinated learning experience that for some may be comprehensive and for others limited. It is almost always the non technical skills that are neglected on the journey along the management road and yet managing and leading people effectively demands superior non technical personnel management skills. The winter 2007 edition of The Journal of the Honourable Company of Master Mariners contained an obituary of one of its members, Sir William Codrington Bt., who, it pointed out, had made the following observation in the 1950's

"It strikes me now that personnel management was a skill never formally taught in the Merchant Navy and that poor personnel management was the cause of much misery and general inefficiency… many…allowed their personal prejudices to blind them, inefficient and unhappy ships were the result."

A large part of the leadership skills that we are discussing are personnel management just as they were in Sir William's time. To equip any employee for advancement into a role that includes responsibility for others, whether it is one person or a large group, it is essential that leadership skills are acquired in the same way as the hard technical skills such as financial management, sales and marketing or engineering. Leadership skills should be incorporated into any formal transition training program. The importance of hard skills is not disputed. They provide stepping stones to success in their own particular field. But they stand alone. In isolation, they are not enough to allow people to realise their full potential as successful business leaders. Without the development of leadership skills alongside hard technical skills success will always be mediocre at best.

The world is full of all sorts of characters; all have strengths and weaknesses, including you. Reflect as you progress through the book, begin to draw a personal picture, consider your own personality and your behaviour, your strengths and those areas that would perhaps benefit from improvement. Consider how they impact on others and how they are perceived by others. People with strong opinions often have a tendency to dominate. Do not be overwhelmed by them, look deeper.

Your task in becoming a leader is to modify your behaviour through proactive learning and then to accelerate your performance onto a higher level while taking your followers with you. We all have the chance to succeed as leaders but only those who apply themselves with compassion and determination will do so.

Part one:
Why leadership?

Chapter 1

Setting the leadership scene

Summary

This chapter provides some background on leadership. It describes the special position it should have in the hearts and minds of officers, managers and supervisors who aspire to be more effective. Key elements of leadership are discussed and in particular the need for it to be sustainable. The role of leadership development as a training function is then considered. Finally the dual concepts of the Leadership Pyramid and the Hand of Leadership are introduced to illustrate why Leadership Throughout is an essential proposition for any progressive work unit.

What is leadership?

Leadership has become an "in" word for managers, human resource specialists, consultants and recruiters alike. They talk of good leadership, bad leadership and the lack of leadership just as a decade ago the talk was of good management, bad management or the lack of management. Today business attributes most failures to bad leadership. Leadership is on a pedestal, seen by many as the pervasive panacea that can cure all business ailments.

Leaders have to work with their followers to produce exceptional performance, leaders alone cannot achieve this. Good leadership is not a solitary occupation. It is an occupation that is facilitated by the leader but only achieved with the consent of his followers.

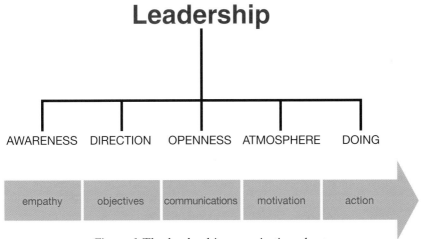

Figure 1 The leadership organisation chart

Figure 1 captures some of the important elements of leadership. These range from establishing and promoting the objective goal or vision to ensuring that words are turned into action. Leadership sets the rules of conduct, it establishes the collective values and creates the mood in which exceptional performance is encouraged and volunteered. Leaders drive change – if there is no change then there is no need for leaders. Change is relative, it is any movement away from the status quo. This might be the capturing of a new market or it may be coping with downsizing when competition begins to overtake a traditional business.

Good leadership is an objective for us all, just as good health is. The reality though is that good health is a consequence of many contributing factors and actions. These may be individual actions such as a regular exercise routine, a healthy diet, regular good quality sleep and not smoking. However external factors, such as genetic inheritance, passive smoking and other environmental intrusions, can prevent us from securing good health. We can educate ourselves to be aware of these potentially damaging factors and then we have the choice of doing things to mitigate them.

Leadership is not so different from this. There are the obvious ways to improve its quality, such as the way we speak to people, the amount of respect we show for others and the messages that we give them. Then there are the external factors that affect the atmosphere that surrounds our leadership endeavours and influence the final outcome. For example there is not going to be much interest in having a supportive discussion with your people if you continue to pay them 40% below the market norm or manning levels are kept so low that it is impossible for tasks to be completed or you fail to provide them with the tools necessary to perform their tasks in a timely and safe manner. To achieve a state of good leadership aspiring leaders have to attend to a wide range of influencing factors.

> ### *Excellence*
> *"In the race for excellence there is no finish line"*
> *H.H. Sheik Mohammed bin Rashid al Maktoum,*
> *Vice President of the UAE and Ruler of Dubai*

Few, if any, of us have a natural disposition to be effective at tackling them all. Some people have particular natural skills in some of them but not in others. These natural skills may for example be the ability to listen well, to empathise effectively, to negotiate and persuade successfully or to be assertive. Most of us do not have all the natural skills we need to become effective leaders, however they can be gained through learning and development. We can acquire the ability to handle wide ranging issues more effectively just as we can acquire new technical skills. To do this requires a level of personal humility which can be seen as the ability to openly admit that we are not perfect and being prepared to learn from others. This enables us to admit privately and sometimes publicly that we can improve our performance. Any aspiring leader can look at his own workplace and quickly identify areas that are preventing his followers from excelling as a group. He can then plan actions to mitigate their impact on performance.

Committing to sustainable leadership

Sustainable leadership success is most likely to come to those who are interested in understanding themselves and in helping others to develop. These potential leaders do not feel threatened by those who work for them or those they work for. They are willing to empower, delegate, coach and guide them in the execution of their jobs, they are happy to develop their own successors. Do you share this degree of engagement with those around you? Are you really willing to develop your own successor? Unless you are able to answer these questions in the affirmative your leadership efforts will not be creating an environment of sustainable leadership.

Sustainable leaders are committed to excellence in the context of an ever changing environment. Their aim is to be at the forefront. Always guiding positive change and ensuring that traditions of the past do not prevent them from seizing the opportunities of the future. They work to achieve this through all the resources at their disposal and this includes their people. They recognise that leadership is not a solitary occupation and understand that its success depends on the performance of both leader and followers. Personal development and the development of others therefore becomes a significant element in their leadership equation.

The role of training

Reading books does not solve all our problems, they do not prepare us in isolation to become good leaders – but they help. Some of us are able to gain a lot from them, others find them tedious and an unattractive medium for the acquisition of new knowledge. Research during the last century clearly identified that individuals vary in the way in which they best acquire knowledge. Some do prefer individual research and book based learning but others prefer to learn by experiment or from experience or observation.

Leadership development needs to be a blended experience if it is to offer benefit to the maximum number of participants in any program. By this I mean it must cover the development needs using a range of approaches such as, on the job training, experiential learning (learning by doing), e-learning, selected reading, modular and distance learning, videos, workshop presentations etc. In this way the varying needs of as many people as possible are covered and the potential for individuals to lose interest is minimized.

Seafarers often have access to video and modular learning programs such as those provide by The Nautical Institute and Videotel. However if mariners are to gain real benefit from leadership development there is a need for greater attention to a wider array of delivery methods. This is a real challenge and as such I would contend that the initial onus is for nautical training establishments to take up the cause of leadership development with the support of the ship owning and operating communities and their various professional bodies. Leadership skills should become a part of the STCW protocols for the future and an important feature of shore based officer training. Perhaps there is a need for a leadership endorsement to be added to the requirements for senior officers, after all we already have gas, oil and chemical endorsements. The flag states that take this notion forward first have the potential to reap competitive benefits as the quality of their fleet operations moves ahead of others through a more cohesive and effective technical and people oriented workforce.

The leadership pyramid

Leadership is not the domain of the most senior people alone, see figure 2. It is a skill that is needed by everyone who has responsibility towards others. This responsibility may be direct, for example, towards immediate reports. In other cases it may be indirect, for example, a middle ranking officer has a responsibility to show leadership towards a junior officer in a different department. The indirect leadership responsibility perhaps seems a little intangible. It is never the less very important. Human beings are keen observers and mimics. We learn to speak by observing and replicating what we see around us. We acquire good and bad habits by watching and copying others. Indirect leadership is therefore impacted by the leader's awareness that others are looking up to him or her (they are senior to them) and likely to copy good and bad habits or at least build a picture of what is acceptable and unacceptable by observing him or her.

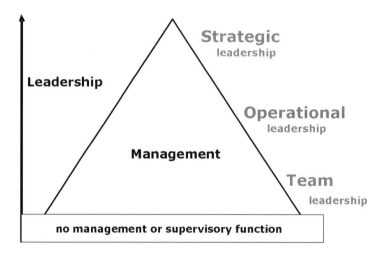

Figure 2 The leadership pyramid

Leadership needs to be present at every level within any organization that wishes to excel whether it is a company, department or division, a charity, a ship or a social organization. This is Leadership Throughout. The higher you rise within a hierarchy the more time you must spend on leadership issues. Nevertheless, as soon as you gain a position where you are even a single step from the bottom of the hierarchical ladder others will look to you for leadership – you are being looked upon as a potential leader, directly and indirectly.

In the mid 1990's I was transferred to Saudi Arabia to take charge of my employer's regional business. I arrived in Jeddah and found our regional head office in the downtown area. The staff were a mixture of British, Indians and Filipinos. I noted that the dress code in the office was very different from that which I had been used to in our UK and European offices. Nobody here was wearing a tie, let alone a jacket. I accepted this difference but nevertheless, I did not dress down myself. I wore a suit and tie everyday, typically British. After a few months I noticed that a growing percentage of the office staff were now wearing ties, some of them had even started to wear cufflinks, as I had always done. I had been observed and others had discreetly changed their habits to more closely align to mine. Leaders are always being watched. Bad habits will be copied just as the good ones are.

Leadership Throughout is a resource that exists within all work units yet only in the most progressive is it exploited in anything other than a subconscious manner. The vast majority of human beings like to know what they are expected to do as well as why they need to do it and where it will take them. By establishing the rules, supporting their people, setting the example and communicating 'the what, the when, the why and the how' of the objective. Managers and supervisors at all levels can begin to exploit the potential of their people through Leadership Throughout. By neglecting this opportunity synergy will be lost and collective performance will never reach sustainable levels of excellence.

The hand of leadership

Throughout the book there are references to The Hand of Leadership (figure 3). The hand has been created from the Leadership Organisation (figure 1). Each finger represents an important component of leadership. We will explore the individual leadership skills required to satisfy each component and look at how they can be developed. Together they are leadership, individually they are not.

> **The hand of leadership**
> *Together the fingers are leadership, individually they are not*

For example it is no use developing a high degree of awareness of yourself and others if you fail to realise the importance and difficulties of giving direction. Similarly being a good communicator and enabling two-way dialogue is no use to a leader if his efforts amount to all words and no action. Each finger has an importance for both leader and follower. Leadership does not exist until all the fingers are in place for both leader and followers.

We will now consider an overview of each of the fingers of The Hand of Leadership. A more in depth discussion of them is contained in the second half of the book.

(i) Awareness – of self and others

There is a famous saying which is used throughout the Arab speaking world to explain differences between people.

"All our fingers are different"

We are all unique, just like our finger prints. The wisdom of the Arab saying is validated every time we see people disagreeing, wherever they are in the world. We disagree because we are all different and we all reach different conclusions.

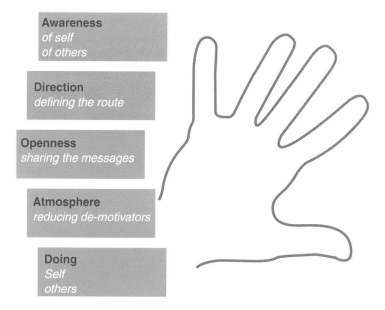

Figure 3 The hand of leadership

Awareness of self and others puts us into a better position to understand how others perceive us and how we all so often reach differing conclusions from what, on the surface, appears to be the same information. To increase self awareness there is a need to reflect on this individual uniqueness and accept it, "as it really is." Unfortunately many people fail to do this, instead they prefer to consider that everyone is the same 'under the skin'. This assumption leads them to believe that they intuitively know how to handle everyone, since everyone ought to feel and understand things in the same way as they do. When the other person does not respond as they predicted they jump to the conclusion that he is wrong or that he has not listened and understood them. Conflict in one form or another ensues. Such an all encompassing approach suggests that there is a human stereotype and this is of course true, we are all *homo sapiens*. But this is a stereotype at the highest level within our species. In reality to treat everyone as if they are of our personal stereotype is doomed to failure, we are all unique and we each have our own unique reality. Just as we do not all like football or cottage pie or hummus we do not all like to be treated in the same way and we do not always come to the same conclusions when we are given the same information. We are all different, just like the fingers on our hands. Given an acceptance of this uniqueness there is a need to explore it and become familiar with its various components. We need to increase our self awareness. Only after we have done this can we hope to move towards an understanding of others and then on to demonstrating real leadership.

Feedback from others is a tool to aid self awareness. Identify people that you trust and talk to them about aspects of your performance and behaviour. Encourage them to be frank with you as you need to know their real opinion, ask them regularly. By gathering feedback, from a number of sources, you will be able to develop a more accurate picture of yourself as a leader. Once you gain a real insight of how others see you it is possible to begin to direct your attention towards an awareness of others. With this new and greater awareness you are well placed to adapt or tune your behaviour to best match your objectives as a leader. Awareness of self and awareness of others is the first finger of leadership. However, alone it is not leadership.

(ii) Direction

Leaders need to start with self direction. Through the awareness of themselves that they have developed they can establish meaningful personal goals and objectives that align with their value systems. Having established these it will be possible to start the process of developing work or business objectives with followers.

Leaders tell people where they are going; they give them the direction. In the absence of direction followers will choose their own path and since we are all unique our paths will differ. If you do not know where you are going any path will do. The lack of direction from bosses, whether they are CEOs, bosuns or team leaders is a major reason for the sub-optimal performance that leads to project or business failure.

The majority of followers appreciate direction. Some like comprehensive direction while others prefer more generic leadership where they have a greater ability to use their own initiative. With this in mind it is of vital importance that as a leader you appreciate how the needs of followers vary from individual to individual as well as from situation to situation.

Dictatorial and autocratic leaders seldom give followers the opportunity to become involved in direction making decisions, instead they prefer to direct on the basis of their own assumptions and knowledge. Such an approach may be okay with unskilled or low skilled workers who accept the transactional nature of their relationship. They are paid to do what they are told and they get on with it. Workers higher up the knowledge tree can, however, find this approach negative. For these people it is unlikely to bring out the best in them. Too much direction can be a significant de-motivator which can lead to resistance and sub-optimal performance. Competent and committed people look for a degree of freedom in what they do and how they do it.

Followers have knowledge and leaders need to have access to it in order to perform effectively. Dictatorial and autocratic leaders like those described above miss out on large portions of this available knowledge, they are not open to the opinions of those below them in the hierarchy. Not only do they put the commitment of their followers at risk, they also deny themselves access to important information and opinion. Failure to establish elements of direction through a reluctance to incorporate the insights of others into their decision making process will in many instances restrict the quality of these decisions. Leaders should take a degree of direction from their followers just as followers do from leaders.

Direction of yourself and others is the second finger of leadership, but on its own it is not leadership.

(iii) Openness

As suggested in the last section leaders can gain a lot from the knowledge of their followers. Open relationships between both can help ensure a strong flow of two way information. Leaders who decide not to share their knowledge and insights with others de-power the potential of their business unit.

Consider the following scenario.

Leadership; a two way process

The superintendent engineer, Mr. Ramesh, was responsible for five identical parcel tankers. The fleet was widely dispersed around the globe, each was assigned two permanent captains and chief engineers. Two of the ships have recently experienced failures of several of their deepwell pumps and Ramesh is having problems arranging for replacement impeller units to be shipped to them from the manufacturer. The manufacturer is adamant that there is nothing fundamentally wrong with the impeller design or manufacture. Ramesh recognises the importance of chief engineers but finds it hard to communicate with some of them. As for the captains, he has little time for them as "they know nothing of our job and most behave like prima donnas".

Ramesh keeps them informed whenever he considers it essential but apart from that he avoids contact. The chief currently on the oldest tanker, Mr Ong, is one of those that Ramesh does not have a good relationship with and he avoids direct contact with him as much as possible. Ong had yet to report any problems with his pumps so Ramesh decided not to let him know about the failures on the other ships and the difficulty he was having arranging spares, it was none of his business. Anyway he thought, as the oldest ship in the fleet any problems they were going to encounter would have already come to the surface; there had been no emergency request for pump spares so all must be OK.

However Ong had been having problems with some of his pumps. He had entered one of the tanks while it was gas free and inspected the impeller on a working pump and on one that had failed. He found that the self lubricating pipe to the impeller was blocked on the failed pump. Ong decided to fit a larger diameter pipe to see if that would alleviate the problem. He got his 2nd engineer to replace the line with some larger bore material that they had onboard. The tests on the pump were successful and the same modification was then carried out on all the pumps as each tank became gas free. Having done that they experienced no further problems with the pumps.

The captain and Ong were pleased with the outcome but decided not to make contact with Ramesh to alert him to the problem and its resolution. Both had a dislike for Ramesh who had always been very critical of their efforts and their ship, they found it impossible to satisfy him. They had frequently discussed his unpleasantness together over the last couple of years. With this in mind they decided that there was nothing to be gained by alerting the superintendent to the problem and their solution, he would become aware in due course. They, of course, had yet to be made aware of the significant failures elsewhere in Ramesh's fleet.

Failures continued on the rest of the fleet with one of the tankers being forced to go off charter for six weeks while spares were sourced. When the superintendent eventually heard of the problem and solution used on the oldest tanker he was furious. In an email to Ong, copied to the captain. he accused him of deliberately holding back information that was vital in order for him in order that he could manage his ships effectively.

Ong, who was approaching retirement, could not resist writing back;

Dear Mr. Ramesh,

Thank you for your recent message.

Over the last three years you have constantly criticised my efforts on your ships. At no time have you ever shared your problems in managing a technically complex fleet with me, or as far as I am aware, with the other chief engineers. It has seemed to me that your objective throughout the period was to marginalise your chief engineers in an errant attempt to feed your own ego. This has made this period very frustrating for me personally and I am saddened to be ending my career in such an environment.

Had you attempted to build a rapport with the team rather than isolating yourself from it I am sure we would have benefited from the collective wisdom that you could have been offered and this would have helped ease your workload. If I had thought that our modification to the deepwell pumps would have been received by you with anything other than your usual contempt I would have advised you at the earliest opportunity. Similarly if you had been more open with us in respect of the problems elsewhere in the fleet we would have been better placed to realize the significance to others of our modifications.

I hope that you will reflect on this letter which is sent in frustration rather than with malice. In your role we needed your leadership. You needed to encourage your followers to be open with you, just as you needed to be open with them. Given free flowing, two-way communications I am sure your job would have been easier, with problems being addressed much more swiftly and effectively.

Yours etc..

In this fictional example there is no winner. The superintendent has failed to mitigate the problem across the fleet in a timely fashion because he did not have the support of his followers, they had not ratified his position as their leader. Similarly while the chief engineer was able to excel in his ability to identify and resolve a problem he failed to share his findings with a wider audience and hence did not benefit the wider team. Had he been recognized as an important team member through appreciation and open communications his last three years of work would have been much more rewarding and his company would also have enjoyed less unnecessary expenditure.

Openness alone does not constitute leadership yet without it leadership cannot exist.

(iv) Atmosphere

Leaders create the atmosphere in which others can excel. Ramesh did not do this and he paid a high price for his failure.

In the late 1950's a American researcher, Frederick Herzberg wrote a book "The Motivation to Work" which reported the results of an extensive research project into workers attitudes to their work (Herzberg et el, 2005). He famously concluded that people were liable to be de-motivated by certain factors at work and motivated by others. People only had a real chance of excelling when both the motivators and the de-motivators were addressed to the satisfaction of the individual involved. The factors were not the same and they each had their own continuums. He called the de-motivators "hygiene factors". The hygiene factors are rather like the lower needs described later in Maslow's Hierarchy of Needs (see page 76) where movement up the hierarchy is only possible once the lower needs are fulfilled. Excessive attention to improving the hygiene factors, Hertzberg asserted, did not lead to more motivation. These de-motivators can be neutralised but any further enhancement beyond this point did not create motivation. Neutralising the de-motivators fulfilled a number of basic needs and once these were satisfied the motivating factors could come into play.

- Company
  - Policies and administration
  - Work and conditions
  - Salary and package
- Relationships
  - with supervisors
  - with peers and subordinates
  - personal life
- Status

Figure 4 Herzberg's hygiene factors

Two examples of the factors Herzberg found are salary (the package) and recognition. He identified "recognition by others" for work well done as a significant motivator of people's attitude towards work. On the other hand he saw the salary that an individual was paid as a hygiene factor, however, his research showed that a perceived disparity of salary between individuals or groups frequently caused de-motivation. He found that the quantum of salary was much less frequently an issue. People were de-motivated when they felt that their salary was unfair in relation to peers or the job they did. Simply giving more salary did not increase motivation (other than perhaps the desire to earn more money!) whereas disparity caused de-motivation. An important point to recognise here is that the issue is, whether the individual thinks the deal is fair rather than whether the boss feels it fair.

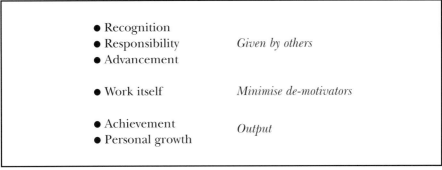

Figure 5 Herzberg's motivating factors

In the UK shipping community seafarers have been fairly well isolated from the problem of salary disparity in that wages are fairly transparently banded against their positions. However the lower wages paid to seafarers from less developed countries creates an international disparity. In such circumstances it is possible to suggest that the disparity may result in lower levels of operational efficiency as a consequence of lower levels of commitment or motivation within fleets where crews are paid rates that are lower than industry norms. In shore based industries there is generally more freedom for managers to find ways to adjust the salaries of individuals in order to resolve specific issues with the person. On occasions job specifications or job titles are adapted to enable managers to "get around the system" and increase the salaries of individuals without the individual's real role being changed. Such actions create the disparities that Herzberg identified. The consequences can be significant for organisations. No matter what endeavours are made by managers to justify the adjustment to other staff members of a similar level, who did not get the rise, it remains the employees rather than their boss who decide if the adjustment is acceptable. If they decide it is not, then de-motivation will become an issue. In most cases, I would suggest that more often than not the employees will reject the adjustment as being unacceptable. That is, unless it is also applied to them.

Leaders have an opportunity to motivate others through recognising their efforts, giving responsibility and by facilitating career advancement. But at the same time they need to focus on ensuring that de-motivators are neutralised. Only then will they have created the atmosphere in which others can excel.

Bosses believe they know how employees should respond to their actions whereas leaders try to understand how their people will perceive and respond to their actions. Leaders pay great attention to reducing de-motivators and ensuring that the motivators are in place. They create the atmosphere in which others excel.

> **Atmosphere**
> *Leaders create the atmosphere in which others excel.*

(v) Action

Leadership is a doing word. Leaders and followers determine their own success by the quality of their actions. All too frequently talk is the order of the day. Lots of meetings, lots of reviews, lots of promises but in the end if there is a failure to convert words to action, leadership has failed. Talk remains a cheap commodity; it becomes valuable when turned into action.

Consider your own ship or office, how frequently do you encounter people who promised action but failed to deliver. To talk is just the start, to deliver is to conclude. Delivering is the essential final finger of the Hand of Leadership. Without action by leaders and followers there is no leadership, but action alone does not constitute leadership.

Chapter 2

The environment and ethics; leadership benchmarks

Summary

This chapter opens a discussion on ethics through an initial reflection on our impact on the environment and the potential consequences of our actions and our inactions. The golden rule of ethics is discussed along with the importance of consistency in our approach to all matters of ethics.

Man's inhumanity to man

No discussion of leadership would be complete without a commentary on the leader's responsibility towards ethical business and his or her consideration for the environment. As we move through the 21st century it is already evident that man's treatment of the environment is becoming a major concern.

The phrase "man's inhumanity to man" has so far been restricted to the description of events such as genocide and the holocaust. Perhaps today we ought to be widening its use to include man's negligence towards his own world. The negligence of current generations towards the environment is going to have an impact on future generations that history may well view as inhumane, a sort of "great grandparents inhumanity towards their great grandchildren" – a miserable legacy by any standard.

Leaders set the example. If they set a bad example many others will still follow them. Leaders need credibility in everything they do and the environment is no exception. Emissions, pollution and waste, for example, are all areas that managers and supervisors can positively influence. With rising environmental concerns, often championed by the younger generations, leaders are going to have to set a better environmental example if they are to remain credible. Those who speak of environmental responsibility, yet do little about it will soon lose the respect and trust of these more environmentally attuned generations.

Fortunately, as a mode of transportation, the maritime industry starts from a relatively strong position. However with 90% of world commodities being moved by sea its total contribution to global warming is huge and it will be in the spot light of environmental lobbies who will undoubtedly continue to strengthen their position in the years ahead. According to a DEFRA[1] report, air transport has a very high climate change impact per tonne *[of commodity]* carried, whereas sea transport is relatively efficient. Friends of the Earth suggest that carriage by sea freight produces one fiftieth the CO_2 emissions per tonne carried when compared with air

---

[1] The Validity of Food Miles as an Indicator of Sustainable Development. Final Report for DEFRA July 2005

freight. It seems that continuing global economic and population growth will ensure the sustainability of sea transport for many decades. If one adds environmental pressures to this, there may actually be a reduction in air freight's small, but significant, share of the global commodity transportation markets.

Leaders of the shipping industry have an opportunity to leverage their environmental responsibility for humanitarian and commercial advantage. Similarly ship's officers can implement best practices through the establishment of environmentally responsible leadership strategies. Leaders influence others and it is within their power to make a difference.

The golden rule of ethics

Ethics is about doing the right thing. However, first we have to know what is right and what is wrong. The golden rule provides a simple and reliable test for ethical issues that we are faced with. It states that we should *"do to others only those things that we would accept others doing to us"*. If you do not like being robbed then your value system should say "I will not rob others." This is an ethical position. Similarly if you feel that others should not wastefully consume air miles you should not consume them yourself. Hypocrites seldom demonstrate real ethics.

We often see "safety" being presented as the most important concern, even ahead of profits and customer service. Yet is this always the case? Or is it just an act of political/commercial correctness to put it on a pedestal while the reality lies beneath it as a current flowing uncontrollably towards profits? If safety is relegated

> **The golden rule**
> Do to others only those things that you would accept others doing to you.

to second place we are acting in a manner that is unethical since the golden rule is potentially being broken.

Consider some of the building sites one sees around the cities of the world. Many look very professional, well maintained, clean and with dangerous areas securely protected. Yet on others you see people working twenty or more floors above the ground in spaces without effective (or sometimes any) barriers protecting them from falling to the ground. Yet all the sites, good and bad, are littered with signs promoting safety. Paying lip service to safety is both unethical and hypocritical. Leaders are failing their people.

From a seafaring perspective consider the dilemma of the ship master which is described below.

A shipmaster's dilemma
Sarah was mid way through her second voyage in command. She was struggling with the pressures she and many others in her position face.

There had been a constant onslaught of weather systems as her ship headed west across the Atlantic towards New York. Now with three hundred miles to go, she was twelve hours behind schedule and fog bound. Visibility was less than one hundred meters. Over the satellite phone, her New York agents had just advised her that she had to arrive at the pilot station within thirteen hours or she would miss her berth.

The ship master's dilemma was that her professional training made it quite clear that in fog she should maintain a speed such that it would be possible stop within half of the available visibility, i.e. fifty meters on this occasion. To do this the ship would have to be almost stationary. On the other hand a quick calculation showed that if she was to make the berthing window suggested by the agent she needed to average twenty three knots for the remainder of the passage, almost her full sea speed.

On her last voyage back to Europe the ship had had engine problems and had encountered dense fog in the western approaches. This led to a late arrival at Tilbury which in turn caused all sorts of problems with the schedule for her other European ports. The office had accepted the problems she had faced, albeit with a certain irritability in their response. Now she was going to be late again.

She made a quick call to her ship manager in head office who had previously commanded the ship. Sounding annoyed, he accused her of being too cautious during the bad weather and of losing too much time. Had she kept up a better speed earlier the delays caused by the fog would have been more manageable. He mentioned that he had often encountered fog at this time of year approaching New York but that it was always patchy and he had never had visibility of less than one hundred meters for more than a few minutes.

The ship master increased speed to fifteen knots. Two hours later she estimated that visibility had improved to almost half a mile. She still felt very uncomfortable; nevertheless, recalling the ship manager's words, she called for full sea speed. Twenty minutes later the lookout screamed that there was a small fishing boat close under the port bow…

Leaders need to support their followers if ethical behaviour is to have a chance. If leaders fail to provide a consistent approach on matters of ethics it is hardly surprising that accidents happen and that people do things that they know they ought not to be doing. Equally important, trust and mutual respect will be at risk, people feel unsupported and used. Someone who has strong ethical standards is unlikely to be an ideal follower for anyone who disregards their ethical responsibilities.

The shipmaster in the situation above was wrong but the mixed messages from head office produced the chain of events that led to catastrophe. What were the ethics of the ship manager? Did he want the ship to arrive in New York on time, accepting any risk to achieve this, especially when responsibility would ultimately rest with the shipmaster? Or did he just use an un-empathetic approach to the problems the shipmaster faced? If the former is true he was being unethical and coercive, which is not leadership. In the latter scenario he was just using lousy leadership.

Chapter 3

Leadership versus management

Summary

In this chapter a generic definition of leadership is suggested and some comparisons between the components of leadership and management are offered. The need for the coexistence of leadership and management within any business or work unit that wishes to excel is discussed. The command and control leadership approach is explained and challenged as an example of an outdated approach to leadership (if it can in fact be called leadership). The case for a greater emphasis on a less hierarchical leadership style is established and finally the chapter links leadership with people.

Defining leadership

Because leadership is defined by the circumstances in which it is practised and by the people who practise it, academics have been unable to reach common agreement on exactly what leadership is. It is then not surprising that it is possible to find hundreds of definitions. However, the one aspect that does have almost common agreement is that leaders cannot lead without followers to follow. There is an interdependence between the follower and leader and only when operating together can leadership exist. Thus leadership is not a solitary activity and the role of the leader is dependent on the led. Followers play a pivotal role in leadership and any discussion on leadership needs to pay proper respect to the significance of followers in allowing leadership to exist. When asked to define leadership today I answer

"[good] leadership is about persuading followers to work together in the most effective manner to achieve the shared [moral] vision."

The words in brackets indicate the implicit understanding I have that leadership must be inherently good and therefore the vision must be inherently moral. This then identifies tyrannical leadership as not in fact being leadership at all. The same argument can be used to say that laissez-faire (could not care less) leadership is not leadership. Anyone that does not care about his responsibility towards others cannot be a leader.

The true power of leadership comes from the synergy of many people working collaboratively towards a common objective. Warren Bennis used the phrase, *"none of us is as smart as all of us,"* to illustrate this point. There is, however, a certain ambiguity about synergy, it appears to defy mathematical logic, how can

$$X + Y = >(X + Y) \ ?$$

Group work is synergistic. The ">" factor is the additional benefit that accrues through collective and collaborative efforts. The ">" factor is what I like to refer to as the latent energy that effective leadership generates. When you experience a negative ">"factor (if that is not an oxymoron!) group initiatives and organisations are put at risk. Positive collaboration in group work creates a spiral of increased effectiveness which allows a group of people to achieve superior results compared to those they would have achieved if they had been acting individually. Leaders have a responsibility to bring people together. They act rather like interference suppressers by allowing the collective energy of the group to focus on a common objective rather than remaining scattered and focused on individual and often diverse objectives. A ship with a tug forward and a tug aft can be held stationary if they pull against each other. Pulling together they move the ship forward. When they encounter strong head seas the two tugs together move the ship faster than one alone, this is synergy.

Without wishing to over write the myriad of available definitions I propose using the definition above as a useful starting point for the purpose of this book. It fits broadly with a substantial number of other definitions and though the semantics of the precise wording may vary the essence is very similar. For example some may say that persuasion suggests too much force or that followers seldom get a chance to really agree with the set objectives. Such points can always be argued but broadly speaking this definition suffices for our needs. Leadership is applicable to any situation where there is leader and follower. It does not matter how big or how small the group is. A one to one relationship often consists of a single leader and a single follower just as a shipping company consists of possibly tens of leaders and hundreds of followers.

How leadership and management coexist

The debate on this subject is very alive but again the approach here is pragmatic rather than unnecessarily complicated. Leadership and management are inextricably linked just like leaders and followers. To be a successful manager is impossible without also being a successful leader and it is unlikely that being a successful leader is possible without a good level of management ability. Management tends to focus on task and process and has a relatively short term orientation. On the other hand leadership focuses on people, direction and sustainable results. Both are essential if a business, or any work group, is to flourish over time. Just as the CEO needs to be a leader so does the first line supervisor. Both have followers, both set an example for others and both create the atmosphere that influences the performance of their followership. Figure 6 illustrates some of the complementary components of each.

The command and control example

In the command and control environment one person usually has a strong hold over those that work for him and he uses this hold to control them in an autocratic fashion. Typically this style of leadership is displayed in emergency situations. For example in times of conflict leaders in the armed forces will display this style and their troops respond accordingly, they are trained and expect to be led in this way. Sometimes a similar approach may be justified in the commercial sector, perhaps in turn around situations where an ailing business needs to be saved from bankruptcy. Unfortunately many leaders depend on the command and control approach when the environment is not so threatening. They issue orders (instructions) and expect unwavering subservience (obedience and compliance). In order to do this they need a high degree of vested power over their followers just as generated in life threatening military conflict.

However, in the commercial arena the power of this style lies in the form of the ability to discipline employees or terminate employment without much justification, or perhaps the ability to prevent career progression. It becomes an almost coercive approach. In such an environment you should not be surprised if you find a subversive response. Performance over time is unlikely to be exceptional.

Management	Leadership
Task	Awareness
Process	Direction
Routine	Openness
Procedures	Atmosphere
Control	Doing

Figure 6 Leadership versus management

Over the centuries shipmasters have been an example of individuals with considerable power. In recent years there has been a move away from the traditional command and control hierarchical structure towards a flatter, inclusive and empowered format as senior managers and leaders have come to realise that individuals alone cannot create an environment in which sustainable group excellence can be expected. If a company, a business unit, a department or a ship is to deliver sustained excellence in today's highly competitive environment it will be because the majority of people involved in the venture are committed to its success. The command and control style does not create such an atmosphere. This poses some challenges for seafarers and the shipping industry. For hundreds of years seafarers have lived and worked in command and control environments. Even today the shipmaster has an almost absolute authority over those onboard his ship. Historically they where referred to as being "master under god" by insurance underwriters and governments alike. The English Marine Insurance Acts of 1906 and 1909 are examples of the use of this powerful title. Given this exalted status it is hardly surprising to hear stories of shipmasters who still wield their authority in such an absolute fashion. Even in the general business community as a whole there remains a core of operators who are still gripped in the arms of command and control management and leadership. People in such situations are often viewed as commodities rather than as valuable assets that can help organisations to succeed or fail.

Command and control environments arguably reflect a style of management rather than of leadership. The followers tend to show signs of being coerced, they do their jobs because they have limited options to do otherwise but they lack the motivation to excel. The focus is on routine and task – the "do as I say" approach rather than any concession towards collective decision making, change and empowerment.

Why do we need leadership?

Successful businesses do not remain successful by luck alone. Whereas it is true that when economies are booming many companies grow and produce strong results the truth remains that many are failing in at least two significant respects. These can provide a meaningful insight into whether the unit is being merely managed or managed and led at the same time.

The first is a failure to keep pace with the changing environment. In a booming economy returns, while impressive, may fail to reflect the true potential of the growing market. The rapid growth in revenues, headcount, investment and profits often mask the fact that neither profit growth or development opportunities are effectively maximised for long term and sustainable success. Mistakes are made in the haste to exploit short term opportunities with the consequent increase in claims and the need for corrective actions. Staff development is limited as there is not enough time to devote to this "luxury". Management becomes stretched and focuses on the urgent, often missing important opportunities. Care in staff recruitment and effective staff development are abandoned. Staff become commoditised and retention becomes an issue. Waste becomes the unspoken reality. The horizon of management draws in and today becomes more important than tomorrow. A command and control approach rules and competitive advantage is eroded. Figure 7 shows how, in order to sustain or improve competitive advantage, a business unit needs to undergo (successful) change at a faster rate than its competitors if it is to stay in front. Without change competitive advantage falters. If there is a small amount of change the loss of competitive advantage continues to increase, only now at a slower rate. The command and control environment is not indicative of change. It is traditional and of limited value in most commercial settings where the need for continuing evolutionary change exists. Management is not about change whereas leadership is. According to Kotter (1996) leadership is the engine that drives change.

The second failing is a loss of vision. Corporate visions (and departmental visions) should give a view of the intended future. However sometimes visions are forgotten (or unknown) and business becomes solely focused on results for today and ambitious growth targets for tomorrow. Fierce competition and a hunger for getting the job done at all cost closes the view and the vision is lost. Once the vision is out of sight things begin to fragment further, the synergy of groups working together in a common direction is lost. Leadership is lost and results are unlikely to be sustained.

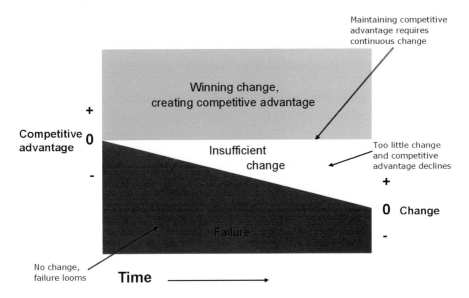

Figure 7 Eroding competitive advantage – leadership drives change

When businesses reach this stage they begin to unravel and competitive advantage goes into decline. Typically the command and control mechanisms get strengthened, a blame culture flourishes, scapegoats abound. No one gets into their helicopter and rises above the frantic day to day environment to see where they are going. This is management without leadership. Given time it will fail just as a flower without water will die.

Leadership harnesses the people

Leaders rise above the day to day and see where the business needs to go. They set the objective and then persuade others to work together to achieve this shared vision of the future. They allow people to develop by ensuring that they have adequate resources, are truly empowered and supported in their endeavours. Bennis (1998) puts it this way "Only a handful of organisations have even begun to tap into their primary resource, their people, much less given them the means to do what they are capable of doing." He continues "…ruthless management may succeed in holding change at bay for a while, but only visionary leadership will ever succeed over time."

Given the need for sustainable success it seems that leadership is a vital skill for managers at all levels. The shipping industry should be no exception. There are well run ships and poorly run ships. It is the well run ships and fleets that have the most sustainable future and leadership at all levels creates this sustainable environment.

Leaders spend time on people and with people. They work at removing barriers, building trust and giving others the freedom to operate without fear of recrimination. The command and control methodology fades and is replaced by a style of leadership focused on mutual respect, partnering and the delivery of long term excellence. In such environments expectations are high but communications and transparency are plentiful. Staff are given the opportunity to develop and a coaching culture pervades the organisation. Once it is seen that the success of the business is being translated into the success of everyone the synergies available begin to rapidly multiply and the organisation quickly accelerates away from its competition.

I have experienced cultural diversity since my late teenage years at sea. The huge diversity in so many workforces seems to me to be both a blessing and a frustration. Often it is used as an excuse to justify the maintenance of a command and control structure, "because that is the sort of treatment people from country x or y best respond to". My question to a person with such thoughts would be "what other approaches have you tried?" to which a likely honest reply would be that they have never tried another approach. In the Middle East one has only to look at the high service levels that are experienced in the local hotel industry to see that people from all cultures can, with proper support and development, operate relatively independently and effectively without the need for rigid control. Diversity should be seen as an opportunity rather than being used as a justification to avoid change.

Leaders thrive on the challenge of change, managers generally prefer the status quo. As Franklin D Roosevelt once said "There are many ways of going forward, but only one way of standing still." Leaders never stand still.

Chapter 4

The born leader myth: False or true?

Summary

Starting with a look at the potential results of perfect leadership this chapter moves on to discuss the stereotypical loud and sometimes charismatic leader. A quote from Harvard Professor Warren Bennis urges us to dispel the myth of the born leader. Good and bad leaders are differentiated. Turnaround specialists are then discussed in the context of sustainable leadership.

The perfect leader

If such a person is to exist, most would suggest that he needs to achieve a number of things which would probably include:-

- exceptional work unit performance which is sustainable
- an inspiring and rewarding workplace
- job security for his or her followers
- meeting the needs of all stakeholders

Delivery of these achievements collectively requires exceptional performance and exceptional skills. It is unlikely that they could be delivered solely through being born with a natural ability to lead or through luck. Things can fall in place through good luck on occasions, but sustainable good luck has to be created by hard work. Leading and leadership is more than simply being able to inspire a certain group to follow you, it requires an active endeavour to do so over time and without unnecessarily disrupting others. Leaders have a moral and social responsibility towards both their employees and stakeholders. The leaders in the Enron collapse of 2001 and at Worldcom in 2002 are examples, on a massive scale, of leaders who ignored their moral responsibility. Our perfect leader has to have an understanding of those with whom s/he impacts, whether directly or indirectly, whether in the short or long term. S/He needs to be highly aware of his/her environment and of the needs of others. He spends his life acquiring knowledge and insights from others. He constantly seeks opportunity for sustainable change and improvement. For such people the age old adage that "change is the only constant" is part of their reality.

Do all leaders have loud voices?

People often form the opinion that he who talks the most, has the loudest voice and shows the greatest confidence is a successful leader. It is easy to identify the leader of a gang but is he necessarily the best leader in the gang? He will almost certainly be the 'largest' character amongst them but is this the best attribute to create sustainable success? Throughout history tyrants have usually been larger than life characters. Whereas they may have achieved short term "success" their legacy is usually one of a flawed and failed leadership rather than anything that is sustainable. Imposing yourself as the leader on others is not leadership, it is dictatorship. Leaders can be made. Being loud and displaying huge outward confidence are not obligatory skills. The noise, aggression and egocentric behaviour that is associated with many stereotypical leaders is not the best model to try and emulate.

Many loud, larger than life, leaders surround themselves with secrecy and use covert methods to hide their weaknesses. In the increasingly competitive environment in which we all operate today they are becoming more and more exposed. What is needed are leaders with a deep awareness of themselves and an understanding of the perceptions of others. Leaders without a deep sense of awareness are finding it harder and harder to maintain their positions. Their bluff is being called and leadership today is expected to be substantive rather than a veneer that hides the true ability and intent of the individual. More subtlety is being introduced to leadership as organisations become less layered and more inclusive. The most successful organisations are sharing leadership throughout their hierarchy, creating opportunities for more junior ranks to assume greater responsibility, allowing them to have a voice and most importantly listening to these voices.

Historically charisma has been seen as a trait of leaders. Charismatic people are able to draw others towards them because of the way they present themselves. They have an aura about them that gives confidence to others. Unfortunately it seems that the confidence they give can be unfounded unless they have the acquired skills of leadership to support their natural talent. Natural talent without acquired leadership skills is not enough for sustained success. Given this fact, charisma is widely considered less important for leadership success today than it was in the past.

Exposing the myth

It is easy to impress others with tales of success and displays of wealth, but does that show the real substance of an individuals' ability to lead? As Warren Bennis says

"the most dangerous leadership myth is that leaders are born – that there is a genetic factor in leadership"

This myth asserts that people simply either have certain charismatic qualities which are required for effective leadership or they do not. That is nonsense; in fact the opposite is true. Those who recognise the opportunity to become leaders can acquire the necessary skills , they do not have to born with the skill set. Many of the louder "leaders" encountered at all levels in organisations are in fact managers rather than leaders and they often manage in a less than effective fashion. Failing organisations are usually over-managed and under led. Tiger Woods is a gifted golfer but that alone was not enough to get him into the world's number one slot. Learning about himself, his technical game and endless practice has given him the opportunity to excel. The same dedication to development is needed by aspiring leaders; we can all improve our game if we work at it.

Leadership is common sense applied to a business in a holistic and consistent fashion. As has been said before, it is not rocket science and nor does it come solely as a birthright. As the complexity of a business increases so too does the skill required to become a good leader. People have always been an important element of organisations. However today, with the intensity of competition we all face and the easy availability of knowledge through the internet and other media, the need to create a working environment where people can really excel together in the agreed direction is more important than ever before. Leaders have to focus on mobilising their resources if they are to outstrip their competitors and ensure the security of themselves and their followers. Good leaders are needed at every level, in every part of organisations in order to achieve this and these good leaders can be made.

> **The leadership adjective**
> *The need is to create and be "good" leaders*

Good or bad leadership?

Simply to lead is not enough. Look at the world's legacy of tyrants; they were leaders, but not leaders worthy of emulation. History has proven this so many times. Some leaders have had a disastrous impact on companies, their employees, countries and occasionally the world. Consider the German leadership in the 1930s or the Ugandan leadership in the 1970s or more recently the corporate leadership at Enron and Worldcom.

It may seem that it is dealing in semantics when insisting on defining a difference between a leader and a good leader. This is not so; the purpose is to demonstrate that there are a large number of leaders who do not fit within the leadership model offered in this book; that is to say they are not good leaders.

It is difficult to decide quickly if a leader is good or bad. In reality considerable time is needed for a leader to make his mark and for the sustainability of his leadership style to be clearly seen. Leadership is about handling people and allowing them to maximise their collective potential, it is not about systems and processes or personal agendas. Good leadership is about getting the best out of every person throughout an organization in order to facilitate the advance towards its vision. There is no simple definition that unambiguously allows you to establish when a leader becomes a poor leader. Nevertheless by studying character traits you can begin to build a picture and identify the differences between positive and negative leadership styles. In turn you can internalise these distinctions and compare them with your own approach to leading others. It is essential to admit and consider your own areas of weakness. This admission represents an initial step in developing the increased self awareness necessary in order to become an effective leader and is the subject of more detailed discussion in chapter 8. Only once self awareness has been well developed such that personal characteristics are accepted in an open and mature manner can any leader or aspiring leader really start helping others to follow the route to self advancement and improvement.

The environment created by effective leadership is positive. Followers believe in the leadership whether it is from the captain, the managing director or the bosun. Words are turned into action. This is not to say that life becomes easy for all involved, it does not. However it means that life has a clear objective and that the focus is on working towards this objective in a consistent fashion without the interference of politics, favouritism or personal agendas.

Good leaders work with others to set, share and promote the common vision and the values. In the process they build strong teams. They encourage everyone to excel, constantly delegating and coaching while creating a "can do" atmosphere of realistic optimism. In doing this they remain in touch with their work environment and staff. They keep up the pressure, always challenging the status quo and driving change. Good leaders know that people are capable of delivering exceptional results given the correct stimuli. They therefore expect a lot from people, they hold them accountable. Yet they always treat people fairly and with dignity knowing that aggression has no place in a leader's toolkit. They recognise and reward success. They listen to others and consider their viewpoints. Good leaders are respected, trusted and successful, just as their followers are.

Openness, a willingness to listen and when appropriate accepting the opinions of others are essential ingredients for effective leaders. They need to be able to openly accept accountability for their actions and be prepared to admit when they are wrong. They do not feel threatened by those working for them or those they work for. They see sharing information and constant communication as obligatory. They work towards the elimination of unnecessary bureaucracy, which they dislike intensely. To them it is noise[2] created by others who hide behind the barriers it creates in the hope of avoiding being moved out of their comfort zones.

A final look at bad leadership helps to put good leadership into context. Bad leadership is demanding and threatening, secretive, autocratic and dominated by bullshit or abdication. It is ultimately unsuccessful. In the business world poor

Supportive *but demanding*	Sensibly ruthless
Enthusiastic	Consistent
Transparent	Competitive
Quick and decisive	Trusted
Empathetic	Communicative
Successful	Inspiring

Figure 8 Good leadership

---

2 Noise – activity not focused on an agreed and meaningful objective

leaders can often bring short-term success to a company. But their style and techniques prevent ongoing sustainability of that initial success. Fear is used as a motivator, but this works only in the short term. Eventually the organisation becomes demoralised, staff turnover rises dramatically and performance falls. Figure 9 lists some of the characteristics of poor leadership.

The turnaround specialist

Consider the way some in which some CEOs move from company to company. They arrive, restructure, cut costs and improve short-term profits and cash. Shareholder value increases as the markets see strengthening results. Then the CEO moves to his next venture. Sometimes they leave and profits then falter as the results of under manning, poor leadership, low morale and under-investment work deep into the structure of the business – an ideal opportunity for a new CEO to enter and restructure again! Others may leave having reshaped the organisation and set it on its way to greatness. These CEOs are leaders, but they are specialists who have a sometimes vital role to play. Their approach to leadership, if they remain with one organisation for a long period of time, is usually not sustainable. They may be called turnaround specialists or company doctors but they are in effect a sort of short term fix, sometimes saving companies but maybe only driving short term blips in shareholder value while at the same time enhancing their own well being.

Unrelentingly ruthless	Blames others
Arrogant and dictatorial	Fails to walk the talk
Politically motivated	(Always) knows best
Full of self importance	Communicates poorly
Unaware of the reality	Does not listen
Short term success	Eventually fails

Figure 9 Poor leadership

> **Assault**
> "You cannot lead by hitting people over the head. That's assault not leadership.' Dwight D Eisenhower

Whether we like it or not organisations do, from time to time, need to be shaken up. Turnaround experts, specialising in the short term fix, have an essential, though painful function to perform. The true sustainable leader has a real passion for the business and its people and wants to stay around to see the fruits of his labour develop. Driving businesses for solely short-term shareholder gain is an inappropriate strategy for any business that is led and/or owned by true sustainable leaders.

Good leadership must result in the development of a business model that is enduring and lasting which meets the long-term objectives of all stakeholders. In creating this model of sustainability the leader creates more leaders. As Ralph Nader, consumer advocate, environmentalist and three times US presidential candidate, said 'I start with the premise that the function of leadership is to produce more leaders, not more followers.' If, as Nader suggests, leaders can be made then the suggestion that all leaders are born is wrong.

By now you should have developed an understanding of why there is a compelling need for more good leaders throughout organisations. Taking this as the norm for the remainder of this book we will now use the word leader or leadership on the understanding that the reference is to a good, positive and sustainable leadership.

Chapter 5

Becoming a leader

Summary

In this chapter the need for more people to be allowed to practice leadership in all industries is raised in the light of increasing competition and the globalisation of knowledge and technical expertise. Against the continuing background of flagging out of tonnage from the merchant fleets of developed countries the need to create a leadership environment at all levels is highlighted. It has the same importance as in any other industry.

Leadership is discussed as an under utilised tool to increase competitive advantage. The need for leaders to listen and liberate the quieter voices in their teams is stressed after which the chapter moves on to look at the perception that managers have of the work attitudes of their staff, using McGregor's XY theory as a base line. It then discusses the leadership voids that are evident in most organisations and how dispensing with old habits and accepting the need for change can contribute significantly towards becoming a leader and ensuring that voids are filled.

Leadership needs in merchant fleets

On a ship it is not the captain and the chief engineer alone who need to be viewed as leaders. The petty officers, cadets, junior and senior officers all, without exception, need to demonstrate good leadership if the ship as a whole is to perform exceptionally. The seafaring industry is not excluded from the need for good leadership practices. Indeed it can be argued that the merchant fleets of developed countries can only expect to survive if they establish a niche for themselves that focuses on both technical and leadership excellence. The developed nation's fleets have taken the technical high ground as seen in, for example, the large chemical carrier and gas tanker fleets they retain. However technical excellence will only secure survival so far, after all, the global village is seeing that technology and other forms of knowledge are now becoming available to an ever wider audience. By embracing leadership excellence the technically leading edge ships will be operated by leading edge officers and crews. Together this two pronged approach towards a sustainable shipping industry will open the gap with less astute competition. It is certainly better to start the change process and open this gap today, before the competitors catch up.

With shore based businesses the situation is not so dissimilar. Competitive pressures are constantly increasing and differentiation has become an important key to sustainable success. Effective leaders recognise this and the development of people becomes a priority. They strive to move their organisation beyond the single pillar of technical excellence towards a more holistic balance of technical and leadership excellence.

The need for more leaders

Here are two more reasons why you must become a leader. First, leadership aids competitive advantage and second, it is an under utilised resource. For competitive advantage to be improved it is a basic premise that every manager and supervisor needs to become a leader. There are several other reasons why competitive advantage needs to be improved. These include the need for stronger financial performance, more repeat business and greater sustainability. Together such improvements provide employees with better security. For this to happen the disruptive noise of people performing tasks which do not align with their unit's objectives needs to be eliminated and this is a key task of leaders. Once aware, aligned and committed to the objective, performance eroding noise will reduce. Followers will be led, their performance potential harnessed and success will be the result. Delivering this scenario remains one of the key issues and problems that business faces today.

Over the last few decades the focus of management development, in most industries, has been on technical competencies. The skills of effective people management have been relegated to the status of "nice to do" rather than "need to do". Sometimes managers think that this is why they have a human resources department, believing that these people "experts" will look after the people management aspects while they can continue with their focus on the technical aspects of getting the job done. This misconception is outdated and if it is allowed to continue competitive advantage will be quickly lost. Today, with the realisation that people are real differentiators, in any competitive environment, organisations of every kind are beginning to understand the value of developing people and leadership skills to a much higher level than was previously the case. It is becoming an important part of the job of every manager and supervisor.

Technical specialists such as engineers, accountants and scientists sometimes find it hard to adjust to this new reality. Some prefer the status quo, particularly when it allows them to avoid the "soft" issues of people management and development. These are people who generally have a left brain dominance. They are most comfortable when dealing with tangible and logical issues. When intuition and emotions come into play they are challenged. On the other hand right brain dominant people have a more natural disposition towards such matters and are less likely to be driven by the detached logic of the specialist. Our brain dominance is not something we can change or something that is "wrong". It does nevertheless represent an important influencer of our character and thus our behaviour. However, armed with an understanding of why we are at first reluctant to tackle certain areas we are better able to face up to the matter and begin to acquire the skills necessary to succeed. Carl Jung, probably the most famous and influential psychologist of the 20th century, strongly advocated the need for people to become versatile in all their brain areas and to avoid allowing their dominant preference(s) to excessively rule everything they do.

Our personal safe havens are our comfort zones, the area in which we feel free of stress and pressure. Habits create these zones and adopting new habits disturbs this routine. Dynamic people continuously push the limits of their comfort zones and frequently move out of them. They are aware that few things are sustainable without change.

Business talks of human assets, human resources and even human capital but in reality what we needed is action not talk and titles. Managers, officers and supervisors need to become leaders. Organisations need to pay real attention to the development of all their people, after all, the people have the power. Those with foresight and the desire for greater success have the chance to take advantage of the opportunity that the development of effective leadership skills presents. The commercial world, in whatever form it takes for you, is inadequately populated by leaders. It is not that the potential is absent, it is that it is allowed to sit dormant by ineffective or misguided existing leadership and management. All industries need to commit to a reorientation of management attention. People need to be treated as resources with valuable potential rather than as robots[3] that are expected to perform a series of tasks assigned to them by those in power without question or imagination.

With the exception of the military in times of conflict people who always operate in a command and control fashion are acting like the slave drivers of the past. They strive to achieve their objectives through the efforts of others but without respect for their needs. Everyone has the ability to go some way to addressing this matter through their own personal development and the application of the leadership skills in their individual workplace. Others who choose to do as they have always done will, at best, get what they have always got; they will be the losers of tomorrow.

To be a good manager or supervisor today you need good and appropriate technical competencies. However alone that is not enough, you must also have good leadership competencies. We have already said that you do not have to be 'a born leader', you can learn to become a leader. Bill George, a former chairman of Medtronic, author and Harvard professor, said in a 2007 interview with Knowledge @ Wharton "Too often we thought of leadership as something you are born with and I reject that idea. Of course you are born with gifts, but you have to develop yourself."

Why though is this notion of needing to become a leader so critical? Leaders are 'can do' people, they are the people with energy, enthusiasm, drive, passion and the most advanced leadership competencies and skills. The energy they possess arises as a consequence of their self confidence and the strong alignment they have with what they are doing and where it is taking them. They take satisfaction from the achievement of those around them such that it becomes a virtuous circle of achievement which, in turn, raises energy levels even further. They know how to make things happen through others. The more "can do" people in the management and supervisory hierarchy the better. Experience helps people become leaders but only those who are prepared to learn from experience can ever

---

3 Robot, from the Czech robota, meaning drudgery or forced labour or slavery.

hope to really make the grade. Leadership skills and competencies can be acquired, just as technical skills and general management skills can be acquired. The problem is that they are often neglected in the process of developing people. Leading is tough and sometimes lonely, but never solitary. It demands constant learning, endless practice and a very high sense of awareness of the total environment. Becoming a leader is a challenge we all face.

Liberating the 'quiet voice'

If you now reflect on your personal position and your aspirations it should be clearer that by trying to enhance your own leadership skills you will improve your contribution to the business as well as your own long term security. In doing this you will also enhance the environment for those working for and with you. Your level of personal satisfaction will increase as will your potential for further growth in your career and personal life. Leadership is not a given, leadership has to be mastered, skill sets have to be acquired. A manager or officer is appointed through his contract of employment, whereas a leader is ratified in his position by his followers, it is not a job title.

Modern business demands achievement at all levels. Those who lack the will or the skills to meet achievement targets will be left behind, their careers will be unfulfilled and their levels of personal satisfaction will be lower as a consequence. By embracing the concept of Leadership Throughout, organisations and work units, of whatever kind, are able to liberate the quiet prisoners throughout their ranks, giving them a real voice which is heard. When this is allowed to happen personal development is actively encouraged and leaders at all levels begin to focus on reducing the de-motivators that exist in every workplace. Together with the newly empowered team members the organisation or business unit can advance to new levels of performance. All will share the benefits of success through facilitating mutual growth and advancement. This becomes a winning formula where everyone benefits and those benefits multiply as the synergies of strong, focused team efforts build.

McGregor's X and Y model

It is often argued that some people do not want to work, they will do their best to avoid it and when they do choose to do it they do it poorly with minimum effort and care. Such people are a problem for any leader and they need to be identified and handled effectively. Fortunately such people only represent a small minority of the workforce. In 1957 at MIT's Sloan School of Management the academic Douglas McGregor made a now famous speech under the title "The Human Side of Enterprise" (Heil et. al. 2000). It was ahead of its time but today his work is recognised as being fundamental in the understanding of how people perform in the workplace and how leadership styles affect this performance. The speech introduced two descriptions of the way in which people in the workplace were viewed by their manager. It is important to appreciate that McGregor was always referring to the manager's perception of people, rather than what the people actually were. In recent years many writers have confused McGregor's theories and suggested that he described techniques and strategies to manage different types of people. McGregor never did this. In fact he wrote "Theory X and Theory Y are not managerial strategies. They are underlying beliefs about the nature of man that influence managers to adopt one strategy rather than another" (McGregor 1967).

There were those managers that viewed people as "X" types. This type of person, the manager reasoned, wants to work as little as possible, lacks ambition, is indifferent to organisational needs and goals and by their nature resistant to change. Managers **who believed** that most people were of the "X" type saw their task as one of strict control and direction. It could almost be said that they viewed their people as incompetent and only being able to perform adequately when they were being constantly and closely monitored.

Others viewed people as being of the "Y" type. These managers **believed** that people were not naturally resistant to organisational change. People naturally wanted responsibility and would work willingly towards the organisations' needs, they were not lazy. McGregor went on to propose that the majority of people in any workplace were "Y" types. However the crux of his theory was that he saw a problem in the way managers (leaders) formed their opinion. Some often treated their people in such a manner that they displayed the characteristics of "X" types; they did not appear to want to work or to help the business unit succeed. This problem was, in his mind, what today we call poor leadership. Managers and leaders were busy de-motivating their workforce because they had preconceived ideas about them.

The concept of Organisational Development (OD) evolved in part as a consequence of McGregor's work. Organisations began to look at their people as essential rather than disposable assets. Leadership is largely about the human side of enterprise and McGregor's coining of the phrase over fifty years ago reveals a remarkably perceptive view on the way the work world would evolve. Leaders have to develop an understanding of their people if they are to be able to utilise their essential assets in an effective manner. The people have the power.

There will always be those who are lazy, and do not care for the organisation no matter how their leaders treat them. A sometimes neglected area of modern day leadership skills teaching is that the these people have to be dealt with. Leadership is tough and from time to time leaders are obliged to sacrifice some people (in an organisational sense) in order to ensure a sustainable future for the majority. However the leader's effectiveness in such cases focuses on his ability to identify the real "X" types as opposed to those displaying "X" characteristics as a consequence of **his own behaviour** or that of other leaders. This reminds us once again of the need for a high degree of self awareness. We have to know how our attitudes were formed before we can fully relate to others.

The demands of business today leave no room for the traits of lethargy and idleness. It is no longer good enough for anyone to turn up at their ship or office and perform their job without effort, care or positive effect. Those who are really of the "X" type represent real liabilities. Maybe some years ago there was apparently room to tolerate such people. However in today's highly competitive environment pressures dictate that every paid employee must produce an effective contribution. Those employees who fail to meet this minimum criteria are becoming more and more of an endangered species. McGregor's "X" types have a limited future. The same applies to managers who believe that the "X" type characteristics displayed by large swathes of their employees are a consequence of the employee's natural tendencies rather than their own ineffectiveness. Such managers need to change their own perception and behaviour. Only then can they begin to realise the "Y" potential that they had previously suppressed in their team.

Today's business world has high expectations of its leaders wherever they are within an organisation. We are operating in demanding and high powered environments. For business units to succeed there is a need for total commitment from all managers and supervisors, including you.

Dispensing with old habits

So far we have seen the need for Leadership Throughout and been offered a justification for producing as many leaders as possible. In life most things that give benefit require effort and this is a truism where leadership development is concerned. Whereas leadership skills can be acquired the acquisition process takes time, dedication, effort and practice. Inevitably, applying new skills means that old habits have to be dropped and new ones have to be learnt; we have to change certain aspects of our behaviour. This regularly means moving outside of our own personal comfort zones. Behavioural scientists have suggested that changing previously fixed habits takes time. New habits have to be deliberately learned and practiced over and over again before they become natural and indelibly inscribed in our subconscious. This process takes weeks or even months, it is not something that can be achieved over night.

Accept the need to change

For anyone to admit that they have imperfections in the way they relate to others requires humility. Those with a disposition towards arrogance often lack humility and frequently consider that, for themselves, leadership development is unnecessary. This is an unfortunate fact; many in leadership positions think that they are beyond personal learning and development in the one area that is absolutely pivotal to their sustained success - leadership. Many people fail to accept that they could do certain things in a better way, they fail to appreciate that their behaviour is not ideal. Until aspiring leaders are able to grasp this fundamental of personal development their attempts to climb the leadership ladder will be thwarted.

I recall my first trip as a chief officer. The ship was a $50000m^3$ LPG carrier and I was an inexperienced 23 year old. The crew were from the UK. We sailed from Rotterdam towards the Arabian Gulf via the Cape. It was a challenging time for me, I had been given a great responsibility at a young age and wanted to do my best.

The crew were "old salts" and I suspect they had quite a problem accepting such a green Chief Officer. The routine was established with the CPO and I having our 0700 meetings on the bridge each day. I was impatient and inevitably I rubbed the CPO up the wrong way. My job became harder and harder, I became more and more adamant that things must be done my way. As we were approaching an off-port stop at Cape Town it became apparent that the relationship between myself and this particular crew was not destined to work. Head Office agreed with the captain's request to do a crew change at Cape Town rather than three weeks later in the Gulf as had been planned. The crew were changed off port. At the same time a message came to me, via the captain, from the office to the effect that "next time it will be more economic to change the chief officer rather than the crew." It was time for me to do a little reflection, to become a little more humble and less arrogant. I needed to start tuning my behaviour and the way I attempted to lead the crew. There where no more mass payoffs mid voyage in the remainder of my career at sea. I had learned an important lesson.

We will never be perfect and the ISO 9000 concept of continuous improvement is therefore fundamental to personal advancement, just as it is to business processes.

Given the opportunity, the vast majority of responsible people enthusiastically embrace the prospect of continuous improvement and the personal growth it can bring. It is for you as a leader to seize this opportunity for yourself. Only then will you move into a position where you can ensure that the people who work for you have a similar opportunity. In starting the process of self evaluation consider yourself as an individual with the realisation that your unique reality will certainly differ from that of others. If you are to be successful in the evaluation you will need to be absolutely honest with yourself, otherwise the effort is wasted and your opportunity to enhance your leadership skills will be lost.

Good leaders see opportunities, they are not necessarily born with the right attributes, but the opportunity to become a leader is there. Will you seize it?

Chapter 6

Creating leaders: A journey of discovery

Summary

This chapter focuses on the tuning of personal behaviour. It starts by considering how bad moods and habits can be passed on to followers and then moves on to personal change. The differences between our personalities and our behaviour are highlighted and the concept of allowable weaknesses is introduced. The flow from our own unique reality to an effective leadership reality through a broadening of perception, understanding and empathy is explained. Next the remarkable frequency of individuals going into battle with each other as they take a rigid position based on their own preconceptions is discussed and alternative ways of handling these collisions of conclusions are offered. Finally the need for organisations to help develop their existing and new leadership potential is proposed.

Confidence and habits

Leaders need confidence, courage, self respect, energy, passion and an ability to empathise with others. Moods are contagious and a leader's mood is particularly contagious. A leader who is always ill-tempered will soon be surrounded by ill-tempered followers. A leader who constantly shouts and threatens his staff will end up with subordinates who adopt the same habits, as Goleman (2006) writes "Moods like colds are contagious."

Fortunately good habits are just as contagious and it therefore seems logical to let them be seen publicly. The way in which you elect to display your mood and the habits you want to be adopted by others is vitally important and requires a high degree of self awareness, but the choice is there for you to make.

If a person in a leadership position has low self esteem their self confidence will also be low. These feelings will soon be picked up by those that are following them. Small things are sensed by followers. If, for example, their manager or supervisor regularly arrives at work or on watch late, the chances are that some of the followers will soon start to adopt a similarly lax approach to timekeeping. Neither boss nor subordinate is now in a position to excel and leadership is lost.

Even confident managers and leaders can have their self esteem knocked from time to time. Perhaps a meeting goes poorly and they fail to articulate their case or maybe a cargo loading operation that they were in charge of does not go smoothly. We all experience such disappointments from time to time. The skill is being able to recognise the effect they are having on us as it occurs and then being able to adapt our behaviour so that our (temporary) bad mood is not picked up by those around us. Skilled leaders bounce back.

Starting to change

It is often hard for people to know where to start with self analysis and personal change. A suggested first step is to admit to yourself that you have imperfections and to record what you think they are. Perhaps you are impatient, a poor listener or inaccessible to your staff. If you really believe that you have no imperfections you are being blinded by your self created unique reality and there is little hope for you as a leader with a sustainable future. The truth is that we all have failings and we can all improve. Those who think they are almost perfect are often the least perfect and yet their high self opinion prevents them from seeing their weaknesses. They will never be good leaders but they may be able to fill the role of an autocrat, a role that should have disappeared with the tea clippers!

Look at a few people that you have difficult relationships with. What is it about them that annoys or challenges you and what is it about you that challenges them? Write down your thoughts. If you are being honest with yourself you will soon realise that many of these annoyances or challenges can be reduced by a change in your behaviour.

I remember, some years ago, a manager that I had difficulty relating to. He ran a division of the business that I was in charge of and it could be said that, from my side at least, we had a personality clash. I did not enjoy spending time with him and I doubted his motives. He gave me the impression that he resented my interventions as his boss. Results from his business were satisfactory but I still lacked confidence in his ability. Reflecting sometime later on this relationship I realised that I had displayed a number of behavioural failings in respect of my leadership of this individual. I was very judgemental and I let my first (negative) impression of him colour all my future dealings with him. I neglected him because we were very different personalities and at the time I was not prepared to face up to the particular challenges he presented me with. I convinced myself that it was too difficult for me to empathise with him. The reality, my failing, was that I did not try! I did not exhibit the leadership qualities necessary to allow this relationship to develop objectively. My failure meant that the relationship stayed in a state of distant limbo. In addition although I had, unjustly, relegated this manager to the category of being ineffective I did not take any corrective action – I lived up to the reality I had created, rather than facing the issue by endeavouring to view his position and attitude with empathy.

Some time after I had moved from that particular role I reflected again on the situation and was able to admit to myself, that my (leadership) behaviour had been very poor. The list of my shortcomings in that situation was extensive but included being judgemental, failing to solicit evidence, allowing personal bias to influence my opinion and an unwillingness to listen. Subsequently, I have been able to devote time towards improving my performance in similar situations.

Being able to recognise and then react to personal failings is pivotal to leadership success. Now answer the questions below and see where you stand.

1. Are you good at listening and absorbing what other people say?   Yes/No
2. Do you give regular constructive feedback to the people who work for you?   Yes/No
3. Do you give praise to others for small things that they do well?   Yes/No
4. Do you have sufficient patience with your people?   Yes/No
5. Do you always communicate your expectations clearly?   Yes/No
6. Are you consistent in how you handle people?   Yes/No

Consider your responses. A "No" response to any of the questions suggests that your performance in that area could probably be improved. Perhaps you are stuck in your comfort zone, reluctant to stretch yourself and encounter change. Maybe you are happy doing things the way you have always done them. If this is the case you should not expect to see improved performance from your team. Maybe you see opportunities as problems. Developed leaders see the reverse, problems are opportunities; they thrive on them and readily except the need to get out of their comfort zones.

Now look at the comments below about the questions posed above. Reflect on them in the light of you first responses.

Of question (1):

- Listening is essential; it gives you information and it shows interest in those prepared to give it to you. Do not expect the respect of followers if you do not have the time to listen to them and really hear what they say. Leaders do not know everything, they value the opinion and knowledge of others

Of question (2)

- People need feedback in order to improve and to remain motivated. Establish a routine of giving regular feedback in real time to your people. Do this and they know where they stand and they will be able to work on their own weakness. Always give feedback supported by evidence and never get personal. Leaders tell people how they are doing, they are interested in developing them.

Of question (3)

- People want to know that you know when they have done good things. Get into the habit of telling them. A little sincere praise goes a long way. Leaders respect and encourage people.

Of question (4)

- Avoid showing too much impatience towards your people. Often we become impatient because we have failed to give people clear instructions of what we expect from them. This is your problem, not their's, so do not blame them. Leaders never blame other people.

Of question (5)

- Linked to question 4, be clear on objectives and communicate them over and over again. Make them simple, explain them thoroughly, get people to para phrase them back to you so you can check their understanding. Leaders set clear and attainable objectives.

Of question (6)

- People need to understand the way their leader reacts. Leaders react consistently to repetitions of similar circumstances.

Adjusting the mask

The way we view ourselves determines what we believe and how we behave. We each see things from a perspective that is unique, our unique perspective. This perspective is informed through our personality which consists of two primary components. First the genes we inherit from our parents and secondly from the environment in which we were socialised as children. How much of our personality comes from each source is the subject of an ongoing academic discussion known as "the nature versus nurture debate". It is sufficient here to consider that an amount comes from each and that by the time someone is a mid teenager their personality is fully formed and unlikely to change significantly during the course of the remainder of their life. It is our behaviour that we can change whenever we have the desire to do so, our personalities are fixed. In the knowledge of this, controlling and tuning our personal behaviour has an important role to play for any leader.

The Greek word persona, meaning mask, has evolved into the English word personality. There is a mask between our personality and our behaviour which we are able to change and modify at will. Sometimes we display our full personality through our behaviour, good aspects as well as the not so good. An alternative that is available to us all is to modify our behaviour in a manner that is entirely driven by the logic of our brains and the situations that we face. This does not always happen naturally, but by actively developing this approach we gain the ability to over ride our emotions and remain in control and gain much more from many of our encounters with others . We do not have to constantly operate in our default (personality based) mode, it is in our power to change. Leadership needs to be more driven by logic rather than by emotion. Behaviour must be effectively controlled, tuned and adapted if we are to constantly get the best out of others. This takes humility and self control and only becomes possible with considerable effort and practice.

It is not always easy to accept that there is a need to tune behaviour. Many remain in a state of perpetual denial. However a more pragmatic and pro-active approach opens the door to improvement and increased personal and group potential. An open attitude gives us the power to understand and admit our individual shortcomings and to be able to start working towards ensuring that they have the minimum possible impact on our leadership performance.

A leader's job is to work with his people in order to get the most out of them as they move towards the common goal(s). It is not to impose his personality on others. Instead it is to use his behaviour to fulfil the intent of achieving the common goal(s) through those around him.

Allowable weaknesses

We tend to automatically fear our own shortcomings and this can lead us to hide from them. We may be comfortable with the status quo and choose sub consciously to remain in a state of constant denial. The fear of our shortcomings emanates from two sources. The first is related to our personal comfort zones. Addressing shortcomings means changing habits and this, as we have already discussed, can be stressful. The second comes from concerns that we may have about revealing these areas of ourselves to others. Perhaps we believe it is damaging to admit to being less than perfect. Wisdom, though, tells us that perfection is a good aspiration. It is one that few reach, but the closer one gets to it the greater the success achieved.

Dr. Meredith Belbin, the originator of the Belbin Team Roles assessment, tells us that we all have weaknesses and that they are in fact a consequence of our strengths! As an example, people of a highly extroverted nature are likely to have great strength in meeting others and facilitating introductions and soliciting an exchange of information. On the other hand they are often not so talented when it comes to routine and detailed administrative tasks. Similarly people with a high degree of social conscience are often labelled as 'good people' and yet their weakness can be that they are often 'too good for their own good!' They put the needs of others before their own needs. By being less judgemental in respect of the words and actions of others it is possible to open up a much broader perspective on how more can be achieved through and with the people around us. Clearly other people are sometimes wrong, just as we are, however good leaders develop a reflective loop in their processing which stops them from jumping to premature conclusions about others. Our individual impression of others is coloured by our own perceptions and beliefs. Leaders become skilled in the use of an empathetic[4] rather than sympathetic approach towards others, they let go of their personal dogma.

By thinking in a more active and reflective way about your "self" and becoming more aware of the full range of attributes of your personality and behaviour you will be opening up your own world of possibilities with all the associated benefits for you and those around you. Personal change is the first element of the leadership development process. Start changing and the iterative process of leadership development will have begun.

Personal strengths and weaknesses

We all have good points and not so good points. The personal reflective thinking that leaders practice helps them see and address the not so good points. It should also bring them to accept their real strengths.

---

4 Empathy – the power of understanding and imaginatively entering into another person's feelings.

Often we admire people who are outwardly successful and wonder what their particular magic is. We may look up to them without the realisation that our own strengths give us potential for our own significant success. The magic of success is within us all. Their advantage is that they have worked and exploited their strengths while at the same time recognised and effectively managed their weaknesses. More often some of us view the achievements of such people as being beyond our grasp. In reality this is far from the truth. Successful people are able to effectively leverage their strengths while at the same time acknowledging and working to ensure that their weaknesses have the minimum possible negative impact on their performance.

> *Speak up!*
> *Be seen, be heard, be visible!*

It may be very easy to admit some weaknesses and yet difficult to admit others. I, for example, have known since school days that my "artistic" ability is very limited and I have always freely admitted this. On the other hand in my early years as a manager I was extremely impatient. Yet for some time I did not see this as my problem. What I saw was that those around me were either stupid or incompetent, not really a recipe for leadership success. I have since come to understand that we are all different and that impatience by one is seen as impetuosity by another.

Leaders can only be made with the help of leaders; it is a truly iterative process. As has already been said the first step to becoming self aware is to understand yourself, to acknowledge your personal weaknesses and your strengths. Only then can you start the process of tuning your behaviour in order to improve its impact on others. It is only once you have developed this enhanced awareness of self that you can possibly begin to help others to become leaders. In acknowledging your strengths and addressing your weaknesses you will be better placed to succeed.

Developing awareness and empathy

Understanding others requires us to take a view of them "as they see themselves" rather than as we choose to see them. If we apply our unique reality to others we see them as stereotypes of ourselves, assuming that they feel, perceive, interpret and understand things as we do. By adopting a leadership reality check crafted by a strong self awareness and a perceptive awareness of others we are able to get closer to really being in their shoes. We can begin to understand how they perceive things and how they feel, particularly in difficult situations. This is having empathy.

In order to get to this position it is necessary to look in your own mirror with clarity and humility. The need is to develop an appreciation of how and why you and others perceive things differently. Given an adequately heightened appreciation and understanding of personal strengths and weakness and the natural differences that exist between all people it is then possible to tune your behaviour towards others. You begin to show greater empathy. This tuning should focus on areas that maximise your positive leadership impact on others. Get this right and you will be seen as a leader by others, get this wrong and you will remain just a manager or supervisor; there is a big difference. By using this process of reflection and change it is possible to establish your Leadership Reality (figure 10). All that remains is to implement it.

We are all different and academics have spent decades studying how we can better understand and manage this challenging issue. Carl Gustav Jung is probably the most influential of the 20th century's behavioural psychologists. His study of the human brain and our personalities has led to the development of tools such as the Myers Briggs Type Indicator (MBTI) and William Shultz's FIRO-B which can be used to build a picture of the personality of an individual. More recently two UK based development practitioners, Michelle McArthur and Keith Nicholson, have designed an interactive leadership discovery tool called Jigsaw@work. Using a jigsaw they take small groups of people through a journey of self discovery where they begin to build an understanding of their own behaviour and personality. After this period of personal discovery the tool is used to build an understanding of the differences that exist across personality types and how it is possible to adapt their own behaviour in order to get the best out of those they encounter. This type of learning experience is an essential ingredient of any initiative to develop awareness and empathy.

> **Difference**
> In order to make people feel the same it is often necessary to treat them differently. Treat everyone the same and the outcome will vary.

Leadership reality

Improving self awareness is where you begin to develop an understanding of your unique reality. Everyone lives in the real world and yet this world is a reality that they have created for themselves, it is unique to them. Theirs is a reality defined by their own experience, their upbringing and its associated socialisation process. The astronomer royal, Professor Martin Rees, talking at a recent graduation ceremony, challenged the audience to question 'what is reality'. He remarked that he, like most academics, is often accused of not operating in the real world and yet, he challenged, "What is the real world? Are for example the city whiz kids in the real world?"

Reality is what we individually see, from the inside, looking outwards. We are all in the real world; it is just that our world realities are different. I for example have an almost total disinterest in football. To the majority of middle aged Englishmen this may seem quite odd. My reality is different from theirs and yet who can honestly say that I am wrong because of this; I am, simply, different, just as they are and just as you are.

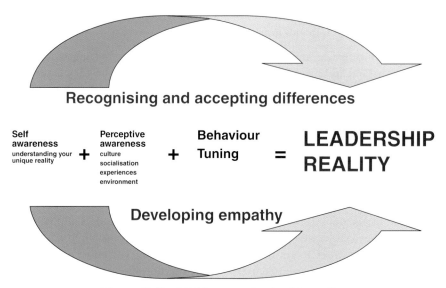

Figure 10 Establishing your leadership reality

The reality for the building labourer from India working in Dubai is poles apart from the reality of the average reader of this book. The Indian worker is probably concerned about being able to repatriate the majority of his earnings home to support his extended family every month. He is eager for every hour of overtime that he can secure and yet he also worries that his company may not pay him on time next month etc. etc. You know your reality, but the chances are that it will be substantially different from the one described above. Our individual realities, our cultural perspectives, our perceptions and our individual values vary. We are all different. It is this difference that makes human beings so interesting and yet the same difference creates the conflict that we are forever encountering, whether it is civil strife somewhere in the world or a disagreement with a colleague at work. Our world thrives and sometimes collides because of the differences of people and yet how much attention to this difference is given when we are learning our professions?

Some of the thought processes that go on in the minds of aspiring leaders need to be analysed, harnessed and probably redirected so that they are able to see the real world of others in a less one sided fashion; they need to develop an effective Leadership Reality. There may be emotional issues or there may be issues of cultural bias or understanding or perhaps personal perceptions. Whatever they may be the aspiring leader needs to become aware of them. They are inevitably things that we automatically grasp tightly, since releasing our hold on them takes us out of our comfort zones and moves us into the unknown. We automatically think that our assumptions are correct. The fact that they are often wrong can be hard to face up to. However we have already seen that the world is full of differences and just because we have a different approach or position on some points to that of other people does not mean we are right and they are wrong. It means we look at things from differing perspectives and that, as a consequence, we draw differing conclusions. In order to determine if there is a right or wrong position on a particular matter we need to view the subject from every possible angle. The aim is to open our minds and adopt an encompassing leadership reality rather than continuing to live in our individual and limiting personal unique realities. The acceptance that your intuitive processes are imperfect requires a high degree of self awareness.

The 'Ladder of Inference'

Stephen Covey in his book, The Seven Habits of Highly Effective People, suggests that in order to get the most out of our interactions with others we should "seek first to understand... then to be understood." How often in your work or personal life have you had misunderstandings which, with the benefit of hindsight, you can look back on and say " I got that wrong"?

Misunderstandings occur for a number of reasons. Perhaps we hear but do not listen or perhaps we make wrong assumption. Perhaps our personal beliefs and values colour our conclusions. Such misunderstandings are more likely to occur in dealings with people from different national backgrounds. Culture adds an additional dimension to work interactions. It brings potential pitfalls but it is not

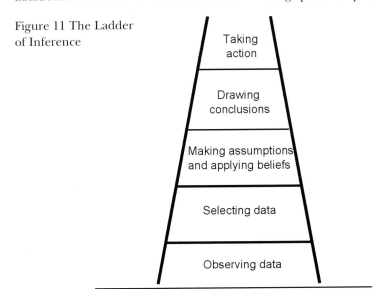

Figure 11 The Ladder of Inference

Adapted from the 5th Discipline Field Book

without its benefits. The merchant service has always been a global business with world wide trading and cosmopolitan crews. By its very nature this has exposed seafarers to a huge diversity of cultures. With the advent of the Global Village (McLuhan, 2001) there has been a dramatic increase in the multi-cultural exposure for people in the developed world, seafarers no longer hold this unique ground. The time when it was possible to spend all one's life working solely alongside colleagues from one's own cultural in-group has passed. In our daily activities it is no longer the case that everyone we encounter will come from our own cultural background. In such a setting the task of being empathetic becomes more of a challenge and is often a major hurdle to individuals as they try to develop what are effectively multi-cultural leadership skills; a must for anyone working in today's shipping or other similarly international environment.

We all have a natural disposition to see things from our own perspective, our unique reality, rather than from the perspective of others. By viewing things through a more inclusive, leadership reality we are able to enter the domain of others with a far greater appreciation of how they need to be handled. Relying on our unique realities is not always the most effective way to have meaningful exchanges with others. We are all different, we all have our favourite hobby, our favourite food and even our favourite people! This is natural. Mankind is made up of differences and without them life would be empty and innovation and change would be absent. The very fact that we are all different is why conflict arises and it is why mutual respect and trust is sometimes damaged or lost. It is not why we argue, but it is why arguments become unpleasant.

Covey's idea of understanding first, before asserting your opinion or dictating action, aims to reduce conflict and disputes through getting people to see the world through the eyes of each other. This is empathy in action. Leaders stand to gain if they can do just this. They need to start trying to see the view point of others and then synthesise this with their own views in order to come to a mutually acceptable outcome.

> **Cultural diversity**
> *The time when it was possible to spend all one's life working solely alongside colleagues from one's own cultural in-group has passed.*

Chris Argyris, a Harvard Professor, introduced the concept of the Ladder of Inference (see figure 11). His ideas were developed further by Peter Senge, of the MIT Sloan School of Management and described in detail in his book "The Fifth Discipline Fieldbook." (Senge et al, 1994) The ladder is used to illustrate how conversations frequently go wrong and to build a picture of how this climb to conflict can be avoided.

Imagine that you and a colleague are in a darkened and sou[nd proofed room] with no distractions. You are both sitting facing a screen and a fi[ve minute video] clip is played to you. After the five minutes you are both asked to ta[ke a pen and] to write down what you have just seen and heard. Comparison at [a later] time will show that you both recollected some items that were the s[ame. It will] also show that you also noted some different things. Perhaps you re[called cars] that were being driven while your colleague recalled the clothes that the actors wore. Perhaps you noted the high speed driving whereas your colleague mentions the picnic hamper in the back of one of the cars. Differences become apparent even though you have both been exposed to the same data. Information overload dictates that we cannot absorb everything we are exposed to and we therefore adopt an unconscious routine of disregarding data that we do not instinctively value in real time.

The above represents the two lower steps on the ladder of inference.

Step 1: Data collection

In this case watching and listening to the video clip.

Step 2: Data selection

The automatic and unconscious rejection of, what we consider to be, unimportant data.

Imagine now that you are climbing one of these ladders while your colleague is climbing another. You are both again in the same room but your ladders are on opposite walls. As you both start to climb your ladders you are moving further apart – the differences between you are becoming greater.

You have now got a certain set of data from the video clip which is, to some extent, different from that of your colleague. Now you can move to the next two steps;

Step 3: Making assumptions and applying beliefs

Step 4: Drawing conclusions

Having completed your data selection you start to "fill the gaps", this is step 3, making assumptions. You naturally, draw assumptions regarding the data, since like all data, it is incomplete. For example you do not know the outcome of the video feature of which you have seen just five minutes. Perhaps because your colleague noted the picnic hamper he assumes that they are on their way to have a picnic. Alternatively you may have assumed that they were likely to have an accident soon as they were driving too fast. So on the basis of personal assumptions one of you expects a picnic while the other an accident – greater difference. You are beginning to tell yourselves a story which is probably light on facts but weighed down with personal assumptions. The area of assumption making is always a dangerous area to enter. We also modify our assumptions because of our own beliefs. Our values and beliefs affect every decision we make. A Christian's belief system allows him to eat pork, a Muslim's does not. Sunshine and good weather to

British person is generally associated with summer whereas to an Arab in the Middle East good weather is more likely to be associated with the overcast skies of winter. Once while working in Dubai I had a meeting with an Iraqi client to discuss some work in Kuwait that I was scheduled to do. As our meeting closed he said "Thanks Richard, I'll leave you go and enjoy the weather now." I was confused, it was raining heavily and I had no intention of going outside to 'enjoy' that weather. However to him it was a novel and pleasant respite from the oppressive heat and constant sunshine that he had encountered for most of his life – our assumptions about the pleasure of rain were very different.

Our beliefs are developed, to a great extent, in the early years of our lives. They relate to our culture, the environment in which we are brought up and socialised in. These beliefs are almost institutionalised in our personalities and as such applying them becomes automatic. We seldom think whether our beliefs will lead us to the same conclusions as others. In fact we often assume that our beliefs are universal beliefs and that anyone who offers a different belief (interpretation) is wrong. This is a source of much friction the world over.

The fourth step on these imaginary ladders is where we move from our beliefs and assumptions and draw conclusions.

The final step is to the top of the ladder where we take action. We either do something or proclaim something. We state our position.

Some time ago I sat in a marina restaurant with my wife, it was very hot. Looking around the marina I noticed a sailor (my assumption!) at the top of the mast of a large yacht. I commented to my wife "look at that fellow working up his mast in this heat, it's a big yacht" to which she replied, "yes it is odd to see someone doing that in this weather but you know the yacht is not that big". My retort to this apparently strange contradiction was "it is big, look it's got three sets of spreaders" to which the reply came "no it has not, it has only got two sets." We then started to explain to each other again what we saw, I maintained that it was a big yacht with three sets of spreaders while my wife stood her ground and insisted it had only two sets. The discussion got a little heated before we realised what was going on. There were two yachts in the marina both with people working at the top of their masts, one large and one small. We were both right. However we had both assumed that the other was focusing on the same piece of data as we were. Assumptions are dangerous. We should each have tried to understand the other person's position before leaping to the top of our ladders and arguing that we were right and they were wrong.

This example illustrates another important paradigm, that it is possible for both parties to be both right and wrong, at the same time. Neither of us was wrong about what we saw (different yachts of different size in different parts of the marina with someone up their masts) yet we were both wrong about what the other person saw. The reality is that the real truth often lies somewhere between the two positions taken – a third truth. A leader is always careful with assumptions and aware of differing belief systems, he is always seeking the third truth that allows both sides of a potential conflict to emerge with an acceptable conclusion.

> **The third truth**
> You draw your own conclusions just as others do.
> Respect their opinions and work with them to find a third
> truth that you can both accept.

No one is perfect but if you gain an awareness that current performance and reaction to events can be improved it will be an early move in the right direction. There is a need to focus on the continued development of "self" as a fundamental prerequisite of future success and happiness. Hard management skills are essential but it is the effective use of leadership skills in our interactions with others that aspiring leaders have to acquire and develop if they are to excel beyond that of being technically competent.

Leadership skills encompass emotional issues and this can be a sensitive area for many people, including those at the top of organisations. However they cannot begin to reflect on performance, behaviour, career advancement, dreams etc. without touching these sensitivities. There is a need to appreciate and accept the challenge that this presents. Once this is done it will be easier to make life changing decisions about what you do and what you want to get out of life as well as how you behave and how you tune your behaviour to get the best out of others. The ability to look objectively at your self in a dispassionate and analytical way is a great discipline and one that you should be keen to practice.

The Ladder of Inference teaches us the folly of jumping to conclusions and encourages us to climb down our individual ladders and see things from the other side.

Organisations and leadership

Organisations everywhere are beginning to recognise the value of leadership skills but few have a clear understanding of what they are or how to maximise them. Unfortunately it is clear from most staff training records that, even today, companies spend the vast majority of their training budgets on hard skills training. What is still being omitted is the recognition that people need to be able to develop themselves to use a wider array of skills in order to be able to lead others successfully. This then provides an effective return on investment as the value of their contribution to their organisation increases through the effective application of leadership skills.

To be the best, an organisation needs not only the best product, it needs the best people:

- with the best productivity
- with the best attitude
- with the best individual and collective leadership skills in the market

I doubt whether any organisation can expect to achieve such utopia if they do not invest in some form of comprehensive training and development in leadership skills for large numbers of their staff. Simply sending senior managers to a high powered business school for a week of "leadership therapy" is unlikely to inspire the rank and file to appreciate, or even sense, an enhanced leadership ethos in the organisation. Ship's officers receive years of training and sit many exams in order to be deemed competent to be left safely in charge on the bridge or engine room of a ship and yet when it comes to managing and leading the crew they usually receive limited planned development. My own recollection from when I was studying for my certificates is that we perhaps spent a couple of lectures learning the protocols for logging errant seafarers but no time whatsoever learning what we could do to try and reduce the incidence of 'loggable' offences. The reality is that everyone who is responsible for at least one other person, whether directly or indirectly, stands to benefit from developing their leadership skills. The aim should be to reduce behaviour that does not add value to the organisation or business unit objective, it should not be to build a mindset that thinks disciplinary processes help organisations to excel.

It is the greater understanding of individual strengths and weaknesses and the ability to communicate, motivate and influence others that helps leaders perform to their full potential. Understanding yourself professionally and emotionally is what matters. Once you understand yourself you can work on improvement and helping others. Organisations need to do more to ensure that their people have the opportunity to explore these areas if they want to leverage the potential synergies that effective leadership throughout can bring. Lee Kuan Yew the first prime minister of Singapore once said "the day I stop changing is the day I start to die". Organisations can suffer the same fate.

> ### *Discipline*
> *Avoid the mindset that says disciplinary procedures help organisations excel. Disciplinary measures are a sign of failure.*

Leadership is all about communicating the goal, explaining why it has been chosen and supporting followers as they move along the path towards it. How the goal is best achieved with the resources available, often involving people from different cultural backgrounds, demands interaction. If effective mutual empathy exists interaction has the possibility of being effective. Without empathy the relationship will develop a one sided perspective and mutual effectiveness will be limited.

In the next chapter we look at leadership styles. Some styles inhibit effective communications while others encourage it. The value of empathy is that it provides a safety net. People who share values will share risks, be alert to danger and work together effectively.

Chapter 7

Leadership styles

Summary

The chapter looks at some of the most significant leadership styles that are seen and written about today. It starts by asking "who makes the best leader?" This is followed by a series of short discussions on the various styles including Hersey and Blanchard's model of situational leadership, an adaptable form of leadership used on a one to one basis. Next the proposition is offered that there is no single 'best' style of leadership and the question "A circle of choice or a continuum of change?" is proposed. The chapter concludes with a discussion on changes in leadership thinking and practices.

Who makes the best leader?

Is it officers, scientists, academics, engineers, creatives or the caring types that make the best leaders? My answer is always the same – it depends!

A creative can be a great leader just as an academic or a very caring person can. However, there are some governing attributes that link leadership success to any of these dominant traits. These are the willingness and the ability of the individual to modify their behaviour to match the situation and context in which they are operating.

We can all modify our behaviour. Just as we have a preferred or dominant personality style it is within us to adjust our behaviour to reflect something that differs from this. Given a willingness to modify our behaviour, the next question is: do we have the ability? Here we can come unstuck. If there is a huge difference between our dominant personality type and the context in which we are asked to operate there may be an insurmountable barrier. As an example imagine trying to get an eccentric musical genius to lead a department full of accounts clerks. Similarly, trying to get a social worker to lead a fast moving, results driven, sales force may not work too well either.

These two examples relate to people in widely differing contexts and situations. Fortunately this is not a challenge that many of us face. Most of us occupy the middle ground, being neither at one extreme nor the other. Given this status and a will to learn to lead we have the prospect of achieving significant results by modifying some elements of our behaviour. It is useful to be able to identify the dominant style of the people we work with. Equipped with this knowledge we can determine how we should behave when we are interacting with them. Jigsaw@work (see page 49) is one tool that I now use to begin to paint a picture of people's dominant styles. A person who is predominantly of an organised nature is likely to have to work hard on their behaviour if they have to work closely with a creative person. The creative person may come up with a myriad of ideas but their ability to follow through and deliver completed projects is probably less certain. The organised person will have to make allowances for this in the relationship. Likewise a dynamic results oriented person may have her work cut out when spending long periods of time with caring people. The carers will usually put their compassion for others ahead of the need for business results yet they have a powerful ability to bring people together and resolve, or at least reduce, conflict; skills that all teams need.

The 'trait model'

This is perhaps the original leadership model which has been seen and discussed for thousands of years. It is sometimes referred to as the "Great Man Theory." Its proposition is that certain people naturally display certain traits that enable them to be accepted as leaders. They are seen as natural leaders and followers readily accept their authority. However research has failed to produce a definitive list of what these traits are though it has been suggested that things such as charisma, integrity, intelligence, motivation, self confidence, energy and flexibility should be considered.

The problem has been that leaders displaying these traits, which may include all or some of those in the list above, are accepted as leaders in any context or situation. It is this acceptance that has been proven to be the flaw in this model.

From another perspective the trait model has a place in very specific circumstances. For example trait identification can be used in the selection process when a major change agent is needed to save a seriously endangered business or organisation. This is the role for the turnaround specialists mentioned in chapter 4. A specific collection of traits that suit the position would in such cases be used in the selection process. Almost certainly three of these would be charisma (to inspire and attract followers), energy (to see that the job gets done) and self confidence (to provide the self assurance that the leader needs to push through the dramatic change necessary). Once the turnaround is seen to be well on the road to success this leader would need to be placed elsewhere, mainly because the often egotistic nature of the turnaround leader does not inspire the sustained collaboration that is an essential part of successful long term change. As Kanter (2000) says "…long marches need ongoing leadership…".

Beer (2001) defines two types of organisational change which he calls Theory O (for organisational) and Theory E (for economical). The former takes time and leads to major cultural change. Conversely the latter focuses on short term shareholder value improvement and is very much a top down driven approach that fits with the trait leadership identified for turnarounds where time is absolutely of the essence and long term impact and sustainability is a secondary consideration. Theory O change rests well with the more collaborative leadership which is discussed later.

Contingency based leadership theories

Moving from the somewhat heroic and blind support associated with the Great Man theory of leadership, where one person is seen to be able to lead, whatever the prevailing circumstances we can move into a more selective group of models. These endeavour to match the leader to the context and situation in which he is required to practice. They maintain that a leader's performance is contingent upon both the specific situation in which leadership is practiced and the specific skills of the leader. This allows natural talent and acquired skills to be matched to the situation while recognising that leadership is not a "one size fits all" skill. I have seen nothing to support the proposition that there is any one style or philosophy that has anything like universal applicability. The objective such as "to save the business" or "to develop a culture that allows continuous change which enhances competitive advantage" coupled to the existing environment sets the context in which a suitable leadership profile can be established and to which candidates can then be matched.

Sir Winston Churchill is a good example. He was of the Great Man theory in his war time leadership role but when there was peace his skill set was not so successful and he lost the support of his followers. Just a few months after the end of World War Two the country removed him as their leader in a general election. Some say that the rhetoric he used after war had ended was too confrontational for the new environment of peace. He had already experienced this type of rejection at least once before. Between the wars he had experienced what has become known as his "wilderness years" during which he was unable to gain political power despite having spent a large part of the World War One as First Lord of the Admiralty with great authority and (generally) much respect. His style excelled when the stakes where high and the country was at risk. In times of peace his skills failed to solicit the same strength of followership.

Situational leadership

In considering the context of leadership one has to reflect on the organisational hierarchy and where the concerned 'leader' sits within it. In that respect the leadership content of the position needs to be considered before any attempt to determine which leadership skills are most appropriate. The leadership pyramid (figure 2) is a representation of how leadership responsibility increases towards the top of the organisation but equally it illustrates that there is a leadership content to the role of every manager and supervisor. It therefore seems that the contention that leadership and management are two distinct but complementary functions, Kotter (2000), is misleading. Does he really say they must be delivered by different people? Mintzberg (1977) rejects the separation of management from leadership, all managers perform leadership functions and leaders that do not manage are isolated and are likely to underperform.

A specific model of situational leadership, developed by Hersey and Blanchard in the 1970s (Hersey et al, 8th Ed. 2000) provides an extremely useful tool for leaders, regardless of their position on the leadership pyramid. Their model advocated the need for leaders to adapt their style in recognition of the prevailing circumstances and in particular the developmental needs of their individual followers. It is a one to one model that offers four styles of leadership that can be utilised when interacting with individual followers; it is not a generic or "one size fits all" model.

The model is built around two leadership behaviours; directive and supportive. When a leader adopts a directive behaviour towards a specific individual he provides structure and organises them through teaching and supervision. This is sometimes referred to as an authoritative style. A sort of "I am telling you what to do, how to do it and when to do it and then I will closely watch you as you do it." In this environment communications tend to be one way. There is limited consultation and the leader makes the decisions. Supportive behaviour is the opposite. The leader adopts a style that focuses on praising, listening, asking, explaining and facilitating. Here the experience is of two way communications and empowerment, a more democratic leadership style.

These two styles, authoritative and democratic, lie at the ends of a continuum and research shows that sustained success is unlikely to result from leaders who remain at either end. Management and leadership usually exist somewhere along the continuum. In situational leadership the position on the continuum varies depending on the particular follower, his competence and his commitment at the time of any 'leadership intervention." The leader decides where he needs to place himself along the continuum for the particular situation, hence the name situational leadership.

Competence, in this model, is the ability and knowledge that the individual has to perform the tasks assigned to him. A new employee for example will typically have low competence by the very fact that he does not know the new organisation well (regardless of any specific technical competence he may have). As another example the competence of an employee starting a new project is usually low as he is unfamiliar with the new task. Commitment represents the individual's desire to do the job well. This can be termed as motivation, confidence and enthusiasm. In this case a new employee is expected to have high commitment as is the person assigned the new task. It would seem reasonable to suggest that an experienced person who loves their work is likely to demonstrate high commitment and high competence in many of the tasks that they perform (but probably not in all of them).

Each of the two leadership behaviours, directive and supportive, are then further defined into four styles of leadership according to the amount of direction and/or support that is being given.

Hersey and Blanchard present these four styles on a chart, see figure 12.

They are clear that there is no one "best" leadership style. Each of the four has a place in the leader's repertoire. A leader may, over time, adopt all four styles for the same person. He may use more than one style on one particular day. Situational leadership therefore requires flexibility from the leader so that he is able to regularly modify his style in recognition of the individual, the situation and the context.

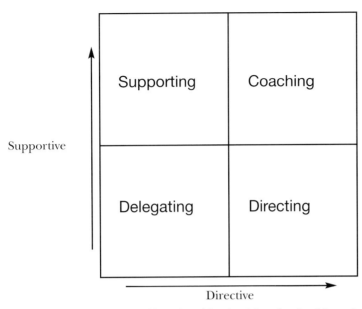

Figure 12 Situational leadership – leadership styles

To use this flexibility effectively the leader needs the ability to diagnose which style is necessary for each situation. In real life we often operate as if we are on autopilot. We intuitively "know" the correct way to manage each person or situation. Or do we? Our intuition is based on our experience and our preferred style. However a more appropriate course of action would be diagnose which style is most appropriate for the current situation. As individuals we have our preferred style but this preference is not appropriate for every situation. If, for example, a person is naturally autocratic with everyone, visualize the relationship he would have with someone who is highly committed and highly competent? The person would soon lose his commitment while his competence would remain intact and his output would drop appreciably. The leader has to become a partner with his people and in this case the need is for a lot less direction in the style adopted. The leader has to become attuned to the leadership needs of the individual and work in partnership with them to achieve long term and sustainable success.

Hersey and Blanchard talk of the developmental status of the individual and view people as having the potential to move from the developing to the developed. As individuals move along this path the leadership styles that will most benefit them vary. The stages of development are referred to as D1 to D4 and illustrated in figure 13.

A person at the D1 stage is high in commitment but has little competence. This does not mean that the individual is a poor performer or the wrong person for the job. The model explains competence as an individual's ability to do a task through their knowledge of the organisation, its systems as well as through their own technical and commercial expertise. The limited competence comes from a lack of familiarity with task and surroundings. This is typical of a new employee or an existing employee taking on a new project or assignment. The commitment is high, perhaps because of the excitement of starting with a new company or having been assigned a new project.

As the individual progresses in his new position it is likely that competence will start to increase (a little) but commitment is likely to decrease as the scope of the task ahead is realised or perhaps as realization that the honeymoon period of the new job is coming to an end. The person is now at the D2 stage. Confidence returns as the new job comes under better control and competence begins to climb, stage D3 is reached. The final stage, D4 is reached when competence and commitment are both high; the person knows what to do, is capable of doing it and is enjoying what he does.

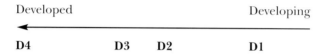

Figure 13 Situational leadership – stages of development

In figure 14 the various leadership styles are referred to as S1 to S4. The selection of which style to use corresponds to the stages of development of the individual. This development relates not to the individual as a whole but to the individual and the specific task that is being considered. Different styles may be required for the same individual on any single day.

It becomes obvious that the leader's task is to assist everyone that works for him to progress along the S1/D1 to S4/D4 continuum so that they become fully developed. At that stage the leader needs to spend very little time with them as they know what has to be done, how to do it and have the commitment to deliver the result. If most people can be led towards the D4 level then the leader's time can be devoted much more towards the needs of newcomers, new assignment takers or people suffering particular difficulties in their work – this produces very effective leadership which is extremely flexible and reflects true partnership with its people.

It is worth considering how they look if we overlay them together on the original chart shown in figure 12 as opposed to the more simplistic view taken in

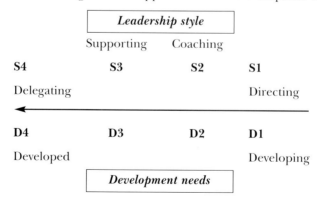

Figure 14 Situational leadership – style and development

figure 14. In overlaying the developmental stages D2 and D3 are omitted since it is possible for them to exist in more than one of the quadrants whereas D1 and D4 are each largely applicable to only one (see figure 15).

The Coaching and Supporting styles differ only in the amount of direction being given. They both offer significant support and encouragement to the individual as he builds both confidence and commitment. In figure 15 the D4 follower is shown as ending with an exit! This is figurative only, but it does illustrate that through successful leadership one is able to develop others for future opportunities, "Good leaders, develop leaders."

I once had a manager in her early 30's working for me. We'll call her Jane. She managed a small laboratory business which was focused in her area of technical excellence. A need arose for us to employ a manager for a much larger and more complex laboratory with a broad range of specialisations, though not in her particular area. External candidates where interviewed with limited potential being identified. Jane was interviewed. She was certainly intelligent and had already gained some experience of laboratory based services. Most importantly she was excited by the challenge. However, the new role would be a big promotion and one that would take her away from her specialized area of activity. I decided to offer her the position and after a couple of days of reflection she accepted.

My task was then to coach her through the early stages of her new appointment. People in new roles are sometimes vulnerable. Hersey and Blanchards situational leadership model illustrates how new starters usually have high levels of commitment (they are excited) but lower levels of competence (they need training and time to become familiar) in respect of the specific position they

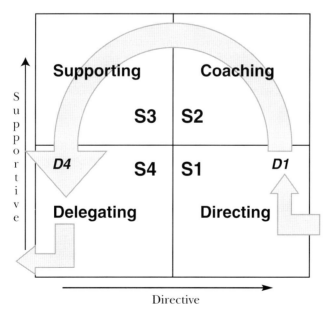

Figure 15 Situational leadership

are assuming. This is not technical competence but competence acquired through familiarization with the new surroundings, the network and the processes. Leaders need to be aware of this vulnerability and be ready to direct and coach the new incumbent as they develop into their new responsibilities. This happened successfully with Jane and very quickly it was possible to move to a supporting role, offering encouragement rather than direction. Jane rose to the challenge and became a proficient manager of the more complex operation.

Had the early support and direction not been there then the results would probably not have been so good. Similarly if had I maintained the directive and coaching approach it is questionable whether Jane would have been able to demonstrate her proficiency so swiftly. Too much directing and coaching would have been likely to have cramped her style and even frustrated her, making her ask the question "why don't they trust me?" Good people need to be empowered, keeping them under too tight a control acts like a constriction in a pipeline; enthusiasm and performance drops.

Distributed leadership

The distributed model sees leadership responsibility spread among large numbers. It exists in an environment of empowerment where individuals readily take responsibility for their own actions. Leaders delegate easily and the rules are clear. In order for this model to work there needs to be high levels of trust and mutual respect. Some would say that distributed leadership is encountered when they see an autocratic or dictatorial leader surrounded by yes men who do his bidding. He has distributed leadership responsibility to his direct team. The problem with this is that these leaders never distribute authority. They rule through a group of mercenaries, their direct team, whose loyalty has been bought, one way or another.

Distributed leadership works well in knowledge worker and routine based organisations that are established and producing good results. However, if results start to falter a change in style may be necessary to bring performance back on track. Similarly in a fast moving environment Distributed Leadership faces problems as integrating change across an organisation requires a broad focus on all the needs of the organisation and this is seldom possible when leadership is so fragmented.

The management of merchant ships sometimes display this form of leadership for long periods. The captain and the chief engineer delegate authority to their department heads (i.e. the chief officer, the 2nd engineer etc) who in turn distribute authority down through their own reports. However when problems are encountered the authority moves back towards the top of the onboard hierarchy.

There is an enforced form of distributed leadership in the relationship between ship owners/operators and their shipmasters. Distance causes distribution. Nevertheless even in such cases high levels of mutual respect and trust are necessary if sustainable good performance is to be maintained across a fleet. In developing the necessary mutual trust it is essential to draw up clear and effective lines of communications – both ways.

Distributed leadership appears an interesting concept but the application and research seems to relate, at the moment primarily, to the education sector. Rather like servant leadership (see page 67) it seems to need an organisation of quite exceptional intent and integrity in order to bring great value. The collective leadership it talks of seems dependent on a superior form of senior leadership where there is both intent and acceptance of the need to allow others to lead themselves. A recent research paper, Klein et al (2005), studied the leadership system for emergency action teams in a major US hospital. The paper described 'rigid hierarchy and dynamic flexibility' suggesting a senior leadership that adapts its level of intervention according to the urgency (life threat) and novelty (learning possibilities) of each situation.

Leadership of the team moves from the most senior team member to the most junior involved physician around the matrix mentioned in the previous sentence. This dynamic process is dependent on the willingness of the senior team member to release authority. This is perhaps the way in which leadership responsibilities move up and down the hierarchy of an effectively led ship. The research, though limited to a very specific time bound and immediate team environment challenges some dominant leadership models and suggests that leadership is a function of the organisation or distinct business unit (work group) as a whole rather than a characteristic of an individual.

The quiet leadership model

Quiet leadership can be very similar to distributed leadership. The most noticeable difference being that the quiet leader may almost go unnoticed. The famous Chinese philosopher who is seen as the founder of Taoism, Lao Tzu or Old Sage is reputed to have said

"A leader is best when people barely know he exists, not so good when people obey and acclaim him, worse when they despise him… But of a good leader who talks little when his work is done, his aim fulfilled, they will say, 'we did it ourselves'."

People will feel that they have the authority to take charge and make decisions that will aid the business unit towards its objective(s). Quiet leaders often see themselves as servants of their people. They do not seek personal recognition from others and instead they gain sufficient recognition from their own knowledge that they are doing their job well. Such leaders see no need to assert their authority as a matter of course, preferring instead to guide and encourage others to excel.

Distributed and quieter leadership models create the collaborative success and in so doing the act of leading becomes a function, or even, responsibility, of many of the parties involved.

Autocratic leadership models

These models see authority and great power vested in the leader. They seldom fit within the context of this book's definition of good leadership (see page 23). Dictatorial leadership is one model of this type. Through positional power and a particular hold over followers these "leaders" are able to assert their authority. Followers will usually be following because they have to, not necessarily because they want to. In a commercial environment, whether it is ashore or at sea, this type of leadership will limit the sustainability of good performance.

The command and control environment discussed earlier is under the umbrella of this type of model. Its effectiveness is limited to times of emergency and massive change. It is not a 21st century leadership style that will bring exceptional performance over time. It is a performance inhibitor rather than an enhancer.

Leaders need to watch out for autocratic leadership beneath them in the hierarchy and see that it is discontinued. The time of individual empires or fiefdoms is past. Twenty first century success relies on the collective efforts of groups of people not on the power of a few individuals. The people have the power.

Transactional leadership

The Transactional model proposes that leaders and followers operate within an unspoken contract. The leader contracts with the follower to perform a particular task for a particular reward. The follower has no obligation other than to deliver the required output. The leader has no desire to develop the follower further as their agreement relates only to the agreed objective. The level of commitment and the expectations of both sides are defined. Machiavelli (2005) warns us to beware of mercenaries. Being followed by mercenaries is the risk that this type of leader faces; there will be little loyalty. He is not going to build a team that excels over time.

Transformational leadership

Transformational leadership takes the concept of mutual expectation and cooperation to a much higher level. Such leaders transform their teams and have the possibility of building excellence. They create a new order in which empowerment, innovation and freedom to act are key characteristics. They create a mood for change and a mood in which others can excel. Transformational leaders relate to the long term and leverage their teams to ensure a sustainable future.

Skills and behavioral models

In these models the behaviour of the leader is considered and their potential for a given position is matched around their ability to display certain competencies and skills. This differs from the trait models where the aura surrounding the person elevates them to the leader's position regardless of the situation and context. Skills and behaviour can be taught and acquired whereas the charismatic nature of many trait model leaders cannot. Using skills and behaviour based criteria there is an opportunity to plan to position people into specific positions that match their competencies, a far wiser approach than relying on charisma alone.

A continuum or a circle of choice?

Many of the styles and philosophies of leadership overlap or form subgroups of others. As an example we talk of charismatic leadership as a style whereas I see it as a sub style within the theory of trait leadership. Similarly the contingency theory forms a sub theory of situational theory and vice versa etc.

Figure 16 shows some of the main theories as part of a circle of leadership. The trait theory is unique in that it is based on the inherent qualities of the individual which supports the perennial question, 'are leaders born or made?' The other theories can, to varying degrees, be learnt. I would contend that as leadership learning grows, the function of leadership selection using a trait based approach will decline. Flexible leaders who have acquired a wide range of leadership skills will come more to the fore. Certain traits will always be important, such as intelligence, integrity and determination but equally important will be the leader's ability to adapt to the situation or context. The trait theory tends to focus on the great man concept of leadership but according to Grint (2005) leadership is so crucial that it must not be left to leaders alone and in this respect the great man concept cannot be allowed to dominate leadership thinking or practice. It seems that most tyrannical leaders could be classified as fitting well within the trait theory, though my own leadership definition excludes them entirely from the leadership family.

Since characteristics (traits) deriving from nature and/or the nurture of socialisation during one's early years are relatively fixed and the use of trait theories in the development of leadership is limited. Northouse (2004) offers the opinion that the trait theories are not particularly helpful for informing leadership development since many traits are fixed psychological structures and it is therefore unlikely that training would have much effect on developing them in people in whom they are not inherently present.

A sub category of the Leader as Follower theory is that of the Servant Leader. It appears that servant leadership as espoused by Greenleaf (1998) seems best fitted to a perfect world. Unfortunately the world is not perfect. Machiavelli, Plato and the daily news all tell us so. Nevertheless the idea that leadership can be seen by some as a mission or duty towards others seems commendable but perhaps for this to occur widely we need a perfect world. Servant leaders can appear in business; but only in the best of organisations where freedom of expression is promoted and where the hierarchy and peers are prepared to listen and adapt their own positions when appropriate. Importantly they appear to need an environment where aggressive conflict towards others is absent. Achieving such an environment may be a noble cause but it will be a hard position to reach. It will only be attainable as one's organisation reaches the higher levels of organisational harmony, synergy and sustained success. This is possibly how Collins (2001) sees the 'Good to Great' organisations. Looking for examples is hard but perhaps WL Gore and their boss-less structure where everyone is called an associate fits nicely, Payne (1998). Since the company was founded by Bill and Vieve Gore in 1958 it has grown to 7000 employees in over 45 locations around the world and has been

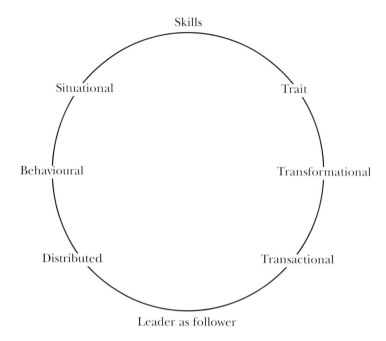

Figure 16 The leadership circle

profitable in every single year. The company remains in private hands with majority ownership now being held by the Associate Stock Ownership Plan which holds stock on behalf of the associates.

To suggest that leadership styles and theories can be presented on a continuum is wrong. To varying degrees all the leadership styles and philosophies have common areas. For example leaders chosen for the specific traits they exhibit will almost certainly have developed specific leadership skills which they can apply in certain circumstances – so trait, skills and situational leadership are drawn into play. Similarly in hard financial times or perhaps on taking up a new position a leader may choose to adopt a transactional style whereas later he may move towards the transformational model – again the leader has used both skills based and situational leadership while acting out transactional and transformational models. It is therefore clear that as long as the definition used for leadership is implicit (that we are talking of good leadership and moral vision) all leadership theories, styles and philosophies can be contained in a circle of choice. The leader then decides on the formula he should use from the available skills and styles to make his leadership most effective for the given circumstances.

Changes in leadership thinking

The charismatic and great leaders will continue to sit in the public spot light and this will always provide them with more fuel for their self promotion. We have discussed Plato's concern that it is the loudest voice that prevails and this remains as disturbingly relevant today as it did two and a half thousand years ago. Everyone has a right to be heard and yet the egotistical charismatics, dictators and autocrats (among others!) seldom have the will to listen or let others take the full credit.

Each of us has a natural predisposition to behave in a certain manner. However it is clear that we also have the ability to modify our behaviour to suit changing situations. There is, I believe, a need to be constantly empathetic towards people and situations and to seek to always adapt our behaviour as the context and situation changes. There is a need to become a multifaceted leader, rather that a monoleader. Personal flexibility is a key to success. Inflexibility is the rod we make for our own backs.

Practicing leadership

Leadership is a skill and skills are transferable. I would suggest that developing leadership skills in others is one of the most powerful weapons we have at our disposal to change businesses and organisations for the better.

There is a case for the rotation of leaders as context changes. As we have already seen, there are those leaders who struggle to adapt well to changing situations and contexts. They are best suited to lead when the situation and context matches their style rather than the other way around. Such rotations can be avoided if the particular leader has been able to develop a skills based style that can be used in a situational fashion. The skill is to adapt one's style to achieve best fit with the situation. The situation that I refer to has three primary elements.

(i) The commercial situation. Leadership in a turnaround situation will be very different from leadership in a stable and routine business environment.

(ii) The hierarchical situation. Senior leaders, those high up the Leadership Pyramid (figure 2, page 10), need a different approach from that of the front line supervisor. Leadership styles have to be adapted as the organisational hierarchy is climbed

(iii) The people situation. People react differently to the same stimuli and leaders have to be able to treat different people differently to get the desired reaction.

The hierarchical context of the particular leader is, in my opinion, of significant importance when one tries to categorise the leadership style that will produce the most favourable results. I am convinced that there is a huge amount of leadership talent in the workplace that is being stifled by the power and authority vested in middle and senior management. The challenge is to find a way of releasing this latent energy for the benefit of all involved by improving the leadership awareness and responsibility of these middle and senior managers. How to do that is the theme of the second part of this book.

Part two: The practice of leadership

The hand of leadership has five fingers

Chapter 8

Awareness: Of self and others

Summary

This chapter introduces the first finger of the Hand of Leadershp. It builds on the journey started in chapter 5 and begins to piece together the Hand of Leadership (figure 3). When complete the Hand of Leadership represents effective leadership in action. In discussing awareness of self and others this chapter looks at behaviour types and then considers the fundamental needs of people. It considers how the behaviour of managers, supervisors, officers and anyone else involved in a leadership function has serious implications for the performance of others. Maslow's Hierarchy of Needs is introduced. Finally behaviour tuning is discussed with a view to improving leadership performance and the performance of others.

The Hand of Leadership

Awareness
of self
of others

Tuning behaviour and using empathy

Direction
defining the route

Openness
sharing the messages

Atmosphere
reducing de-motivators

Doing
Self
others

Behaviour types

Typically we have the capacity to display three types of generic behaviour, assertive, passive or aggressive. Which one we use depends on our personal disposition and the particular context and situation in which we find ourselves. It is usual for each of us to have a natural preference for one or other of these styles. However, as leaders we need to be able to control our style and remain within the limits of assertive behaviour, if we are to maximise our effectiveness.

Looking at how the three types of behaviour are commonly understood we see a clear indication of the potential pitfalls of aggressive and passive behaviour while the potential for assertive behaviour becomes clear.

Passive people are seen as compliant, docile and submissive. They either like to please others or they cannot be bothered to challenge. As a consequence they lack the ability to influence the decision making process or to add value to discussions. They often become the victims of aggressors.

Aggressive people typically display an arrogant or threatening attitude. They are, hostile, strong minded and generally poor listeners. They like to win at all cost. They demand performance through threats and fear. Their style is independent and self dependent; teams are not necessary other than to do as they are told! Aggressive people soon lose the respect of those around them.

> ***Avoid the "big stick"***
> *Instead recognise, respect, coach, and encourage.*

Assertive people are confident, open, strong and respected. Their attributes enable them to be the best influencers of others. They have an ability to state their case while at the same time being willing to listen to others. If necessary they adapt their own needs in the light of input from others. They are always prepared to honestly address issues.

By frequently observing the behaviour of others around us it is possible to recognise the different styles mentioned above. You will see the aggressor who forces his will on others "It is not for you to challenge me. Just do what I say…or else." Then there are the passive types who appear to accept everything that they are told "OK, what ever you say." Finally you will see the assertive types who are able to state their position and what they want done in a compelling fashion without the flaws of being aggressive or passive "I want you to do this task, it is important for the business unit success and I know you have the skills to do it well. What do you think?"

Sometimes by looking at others we can come to gain a clearer understanding of how our own behaviour is perceived. It is not enough to merely recognise and categorise our peers and colleagues, it is essential to understand ourselves as well. The aim in all interactions at work is to create a "win/win" situation. The desire should be to achieve an outcome that is beneficial for all parties. Sometimes this is not possible, but more often than not it is.

If you think you are already appropriately assertive in all your encounters, think again as you are probably denying reality. A problem that many of us face is our own inflexibility to accept that some of our personal attributes are less than perfect. We all have some shortcomings and these may surface in our interactions with others. This is particularly common when there is something that we consider significant, at risk. It is tempting to think we already have all the skills necessary to effect successful negotiations and that we always communicate well, even when something significant is at risk. If this fits with your personal perception of yourself, be careful, you may be more aggressive in your approach than you realise. The line between assertion and aggression is thin and it is important to realise the damage that can be done to relationships and performance by crossing it.

Often, those who do tend to act aggressively, especially when challenged, find it difficult to accept their own negative behaviour. They find it difficult to accept the views of others, particularly those of passive people whose views they develop a habit of ignoring.

Look at the questions below and answer them to yourself as honestly as you can.

(i) Do you regularly take note of the view of others?

(ii) Do you listen to the quieter people you encounter?

(iii) Are you a team player?

Do you feel comfortable about your answers, or are your private responses different from those you would give publicly? You cannot be a team player as well as an aggressor, just as you cannot be a true leader while you remain in denial of the impact that your real self has on others. Aggressive behaviour destroys trust and mutual respect. It leads to a self imposed isolation, where the leader's endeavours depend on power, position and authority. These attributes are, however, not indicative of leadership. They are positional, i.e. they were given to you when you were appointed to your job. Aggressive leaders abuse their positional authority and manage through fear. Assertive leaders respect their authority but avoid using it as the big stick to get things done. Everyone has the right to an opinion, feelings and emotions and to express them appropriately just as they have the right to make occasional, non repetitive mistakes. Everyone has the right to be treated with respect and dignity. Assertive people recognise this and they make the best sustainable leaders.

The success of business units depends on several key behavioural factors, one of which is that its leaders (at all levels) have the ability and willingness to coach others and help them to develop their talents. People with a tendency towards aggression do not make good coaches; assertive people do. It may be difficult for those readers who have an aggressive tendency, to accept the need for a win/win solution, when win/lose serves to bring you some success. The difference is that win/lose is a short term tactic whereas win/win is much more durable, there is no alienation from the other person or people.

There is no place for aggression in a great organisation, it does not foster good relationships or team spirit, instead it creates resentment and destructive tension.

It may result in short term gains but it will result in long term losses. An assertive person effectively influences, listens, and negotiates. Others are then more likely to cooperate and a sustainable relationship that brings mutual success can develop.

Passive people are going to be losers unless they modify their behaviour and show a more assertive approach to others. If they believe that being passive is the only way to behave they will always be victims, often ignored and sometimes abused. Their contribution will be limited and their ideas and opinions will go unnoticed. There is no such thing as a successful and passive leader or manager. Leaders are active in the creation of their own image (Grint, 1997). Passive people create the wrong image.

Maslow's hierarchy

Our performance is seriously affected by how our personal needs are satisfied. Reflect for a moment on how you react when someone, who has significant authority over you, is aggressive towards you. Do you feel annoyed? Do you feel resentment? Do you feel a desire for revenge? If you answered "yes" to any or all of these questions, now think about your work performance in the short term. Logic suggests that your performance would dip, you would become preoccupied with your feelings and perhaps with the desire for revenge. If the aggressive behaviour continued then your performance would become permanently depressed and your engagement with your responsibilities would weaken. Excellence would become an impossibility.

People are driven to fulfil their needs. The behaviour of leaders can have either a negative impact, as in the previous paragraph, or a positive impact on this process and therefore on individual performance. If leaders can behave in a style which allows followers to fulfil their needs then a great latent potential will be harnessed. This is the potential that all leaders should be dreaming of exploiting – getting others to do what you want them to do, in the best possible way, because they want to do it.

> **The performance ideal**
> Getting others to do what you want them to do
> because they want to do it.

Just as we have needs that have to be satisfied in order to survive we have needs that have to be satisfied in order to excel. Imagine yourself in "survival mode". You can only have one of the following three needs; air, water or food. Which would you choose? Now the situation has improved and you can have any two of the three needs, which would you chose?

It usually seems obvious that if we can have only one, we will select air since without air we suffocate in minutes – it is our primary need. Next you probably chose water, since thirst will kill before hunger does.

In the 1940s the American behavioural psychologist Abraham Maslow conducted some extensive research into the needs of human beings, how they were fulfilled and the significance of this in respect to their performance. His work, sometimes referred to as Maslow's Hierarchy of Needs (figure 17), represents one of the most important models of how people's performance is affected by their environment and their current situation. Its findings are used extensively today in the training of many progressive occupations where the need for advanced skills in dealing with others is attributed significant importance. Doctors, nurses and school teachers are among those who are introduced to Maslow's concepts at an early stage in their careers.

As managers and leaders it is very helpful to have an understanding of Maslow's model in order to be able to appreciate how others react when you interact with them. Maslow contended that we have to satisfy our needs in a specific order just as is illustrated in the air, water, food example. He suggested that higher needs could not be satisfied until the lower needs had first been satisfied. e.g. there is no need for food if there is no air.

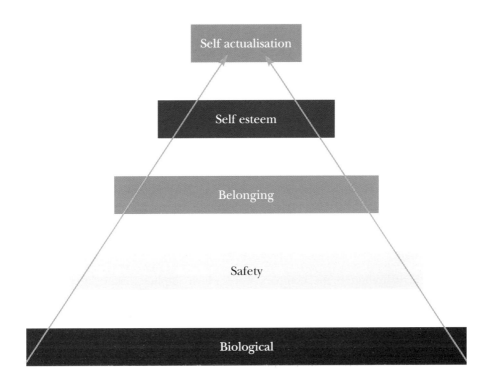

Figure 17 Maslow's Hierarchy of Needs

LEADERSHIP 77

The first, or lowest, set of needs are those relating to biological factors such as the need for air (to breathe), water (to satisfy our thirst), food (to satisfy our hunger). Maslow argues that until our biological needs are satisfied we will have no interest in looking at higher needs.

After biological needs comes the need for safety. The need to feel secure and free from life-threatening danger. Reflecting again on the ordering of needs it becomes probable that in order to satisfy our biological needs we may forego the need for safety. As an example when suffering from extreme thirst or hunger we are prepared to take significant risks to satisfy these needs. Translating this model to the workplace suggests that things such as job security fall into the realms of safety. Threatening an individual's livelihood is therefore more likely to cap their performance at a relatively low level rather than motivating them to excel. Threats are generally a counter-productive tool that are usually not included in a leader's toolbox. Consider the case of the school child who is being tormented by a bully. There is no physical violence but plenty of unpleasant threats. What does the child feel and how does he react? Probably he feels unhappy, maybe he tries to avoid the bully whenever he can and maybe he feels a need for revenge; getting his own back. When bullying happens in the work place the same feelings and reactions are likely to occur. Performance in such circumstances is unlikely to be good as too much energy is being consumed handling the threats and avoiding the bully.

The third level is belonging. People are social animals and as such they have a need to associate with others. They want to be part of a group which may for example be work colleagues, shipmates, friends or members of a club. The need for belonging is however subordinated to the need for safety. The leader who attempts to use threats to get people to deliver results distances himself from them, damages their safety and makes them feel that they do not belong. Once again we have a recipe for underperformance.

The penultimate level is the need for self esteem. People need appreciation from others and a respect for themselves. We each have a need to be valued and in some way admired by others. This may be apparent in a need for power, authority, recognition or even fame. It is when others recognise us that our own levels of self esteem grows, confidence in one's own ability increases and performance leaps towards the fifth and final level in Maslow's Hierarchy of Needs, self actualisation. A good degree of self esteem makes people feel good, when we fell good we perform well.

Self actualization is the complex word which implies a state in which an individual is capable of reaching maximum potential. When all the subordinate needs are fulfilled the individual is in a position to "do what he is here to do". We are said to have reached the level of needs where we are operating without threat, respected and appreciated by others and at peace with ourselves. At this stage all the barriers to performance have been removed and we can excel at what we are best at and move towards an objective worthy of our aspirations. It is also an aspiration that we should allow and help others to have.

Most of us have reached a self actualised state at some moments of our life. My last few years at sea were spent as a Chief Officer on LPG carriers and I can recall one particularly busy period on a 15000 tonne ship carrying ammonia. We were trading between Odessa and nearby Turkish Ports, short sea passages with lots of cargo work, pilotage and standbys. The ship was more than 15 years old and there were always challenges. However on this particular voyage we had experienced no down time and our loading had been completed perfectly. I was extremely tired when we finally let go from the load port but nevertheless I had a 'top of the world' feeling. I was feeling good about my own performance as well as that of the crew and the ship; we had done a good job. I had reached a state of self actualisation, albeit, for a fleeting moment!

In a work context it is useful to understand Maslow's model and to recognise when one is impacting on the needs of others. It helps us to understand how people should be handled and how the mishandling of people limits performance. As an example, the threat to safety posed by imminent redundancies can cause people to lose self esteem and hence their sense of belonging. Their performance falls and their vulnerability increases. It therefore seems logical to ensure that during such time the period of doubt for individuals is minimized as much as possible. The issue of using job security as a threat, on a one to one basis, is not a tool to encourage enhanced performance. Confrontation with people also has a negative impact and Maslow's hierarchy illustrates why and how. This is not to say that all confrontation is to be avoided. Since we are all different, confrontation is inevitable, but it has to be effectively managed. Leaders have to respect their followers and turn confrontation into dialogue where they listen and use the experience gained to make the best collective decisions.

Safety, self esteem and belonging can all be harmed and as the individual descends further down the hierarchy performance collapses. Their focus turns from the benefit of all to the survival of "me." Whereas, when an individual reaches the area of self actualization their performance is optimal. When they are operating in the realms of their biological and safety needs they act in an animal like or instinctive fashion. This is usually represented by absolute attention to ones self rather than to others. In a business context this is shown by an absence of team work, humility and mutual respect and trust.

The areas of belonging, self esteem and self actualization provide managers and leaders with a fertile ground which they can utilize to maximize an individuals personal development as well as his contribution to the success of the organisation.

Behaviour change

With the introduction of the dual concepts that we are all both different and imperfect it should be easy to accept the need for personal behaviour tuning. A book cannot tell you which part of you it is that can best be tuned but it can open your mind to the opportunity and the possibilities presented. When we accept the legitimacy of the Ladder of Inference (page 64) and of Maslow's Hierarchy of Needs we begin to appreciate the significance of the impact that our behaviour can have on others. As leaders this consideration must be mandatory. Without meaningful behavioural tuning leadership effectiveness is not going to be improved. However, once started, behavioural tuning need never end. With each successful change comes an enhanced awareness which enables further tuning; it is another iterative process.

The short list below highlights some areas of personal behaviour that can negatively affect a leader's impact on others, there are many more. Study it objectively and decide if it mentions some areas that you should be working on.

(i) Impatience: expecting everyone to understand everything you say, immediately you say it

(ii) Communications: expecting others to be able to read your mind and make the same assumption as you do or failing to communicate the objective(s) that everyone should be working toward

(iii) Style: being reluctant to adapt your leadership style when circumstances and contexts differ

(iv) Inconsistent: having an unpredictable approach when circumstances and context have not changed

Personal change is personal and the hardest part is to establish the initial momentum. Once movement is started it becomes easier and the fruits of your labour will gradually appear. It is of no use whatsoever to proceed to the act of attempting to lead others to excellence until you have understood yourself and begun to adjust your own behaviour where it is necessary. Too often the need for personal change and honest understanding is denied. Such change has to be learnt and it can be usefully achieved by setting goals for personal improvement. In this way you are able to work on the negative beliefs associated with inappropriate behaviour and overcome them. This is the subject of the next chapter.

Chapter 9

Direction: Goals and values

Summary

This chapter introduces the second finger of the Hand of Leadership, direction. The discussion turns to look at how leaders give direction to others and how they establish their goals and share them effectively with their followers. It covers the problems that inevitably arise when people are expected to operate in a vacuum without any clear understanding of the direction which they are supposed to be heading in. Personal and business values are discussed and their importance in the ethos of effective organisations. The link between managing ones time effectively and values/objectives is established.

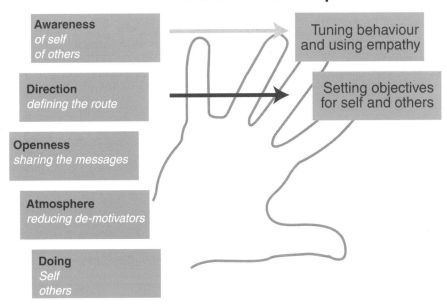

Giving direction with guidance

Leaders guide their followers by offering support and direction; they endeavour to give work a meaningful purpose for all involved. They utilise the situational leadership model discussed in chapter 7, adjusting their style of interaction with people according to the situation and the developmental stage of the individual.

Guiding others involves telling them where you want them to go. This is the stuff of vision and objectives; collectively we will call them goals. In order to reach the goals for their area of responsibility leaders spend much of their time conveying them to their followers. This may be in the form of a CEO, director or senior manager sharing the corporate vision with his staff or perhaps the supervisor discussing team objectives covering the shift for the next few days.

The leadership pyramid (see figure 2, page 10) illustrates clearly how most staff within organisations have some degree of leadership responsibility and this includes the need to share goals with their followers. The subject of goals is then not just about the CEO and the corporate vision, it is equally about the hundreds of objectives that people within organisations have to pursue daily. It is important within organisations for there to be an alignment of goals to meet the corporate vision. At this highest, almost holistic level, good alignment can contribute significantly to a healthy corporate culture where people understand where the organisation wishes to head and what role they have to play to help make the journey a success.

For our purposes we can consider visions as being of two types

(i) The high level goals that come from head offices. These are usually in the form of the corporate vision that ought to guide and shape the future direction of the business in the years ahead.

(ii) The business unit visions, established by leaders throughout an organisation's hierarchy. These must align with the intent of the corporate vision but they can be much more specific to the needs of the particular business unit or department.

On the other hand objectives are the bread and butter goals that we encounter and work with every day, in every thing we do. They may be as simple as arriving at a meeting on time or completing a report by the deadline, completing cargo operations before the laydays expire or personal objectives relating to career development.

Work without goals

What is the purpose of anything we do if it does not have a goal of some sort? People need goals, both in their personal and work lives. Perhaps you have set yourself the individual goal of getting a particular certificate of competency or to securing your own command or paying off your mortgage. Maybe you want to have the tidiest engine room in the fleet or to get your ship or business unit through its next external audit without any major non conformances. Whatever they may be, goals are present every day and in every thing we do that is meaningful. As a leader, establishing a goal and then sharing it with those whose help you need to achieve it, gives purpose. For your followers, it switches on the light at the end of the tunnel; it gives them a target to work towards.

If the goal is missing or not shared with those who will be able to help you realise it, there will be no success; it will be no more than a missed opportunity. Sometimes the sharing process fails, perhaps because the goal is poorly defined. As an example a manager may declare "We must improve our performance". What he means by this is open to individual interpretation. Perhaps he means that costs are too high, or does it refer to poor timekeeping or maybe it relates to preventative maintenance that has fallen behind schedule? Objectives must be clear and unambiguous, they then become something that people can be made accountable for.

As seen in the section above, a work unit without a structured and well focused sense of purpose is not going to be the most efficient. The same can be said of individuals. Personal goals are needed so that we can keep ourselves on track. Working towards portraying an assertive and confident character could be one such goal. However, goals need to be specific if they are going to be really useful. It would therefore perhaps be better to establish a series of more objective goals that together build towards the ultimate goal suggested above. A useful starting goal for a shy or passive type of person could, for example, be "to introduce myself to at least five new business or work contacts in the next seven days."

Corporate and departmental visions

The skills needed for successful leadership and management can only be applied to effect if you know where you are and where you want to go. This is the case whether you are in charge of a company, a division, a department or any other group. If you are lost, your followers will be suffering the same fate. Any direction they take will do; but do not expect to get anywhere in particular or to get there together.

Once you have decided where you are, you can decide where you want to go and establish your vision. A common failing is that this goal is not shared with an adequate degree of clarity. It may be unknown or unclear or misinterpreted by many of your followers. If this is the case it is a bit like walking with a blindfold on. People will go up the wrong path or bump into insurmountable obstacles or just stay still, avoiding the unknown.

As a starting point for a successful journey the ultimate goal must be widely communicated, it should be encapsulated in the company's vision and this must be clearly understood by every single employee. It must be factual and specific. A business will only generate cynicism among its people if its declared vision is vague, unclear or at its worst confusing. The ideal vision statement is short and concise. Some examples:

- To be the best in terms of customer service and profit in every business in which we operate.

- To create an environment in which satisfied clients, quality products and profit go hand in hand.

- To excel in the services provided by the company as measured by our client satisfaction.

- To double in size in 4 years through the development of excellence in our service to current and new customers.

- To achieve personal satisfaction and financial success through the delivery of superior leadership development services to companies in the region.

For a shipmaster, chief engineer or any other professional working in an operational environment the vision embraces the company's sense of purpose. For seafarers this can encompass things such as the successful execution of charters, excellence in ship operations and safety, reliable service and the need to protect the environment.

Each of the statements above is unique and the message they contain is clear. Each employee can see what the company's agenda is. That is not to say that they are all perfect or even necessarily correct. The hypothesis is that the concept of sharing an organisation's ultimate goal with all employees is an essential starting point for every leader. The power of the vision statement should focus and energise the whole organisation and drive everyone to the common goal.

The corporate vision should be in the minds of all staff and so should the department's ultimate objective.

Many companies seem to produce vision statements to appease shareholders rather than to define their objective and focus their workforce. They look good in the Annual Report and in the head office foyer. Such statements will become worthless adornments. More often than not they are forgotten or unknown by the workforce. Corporate visions are sometimes misleading wordy statements that fail to inspire the most important people, the followers. What is needed is a vision that is clear and understood by all of the people, one that they can accept and work willingly towards.

Do you know and understand your company's vision, does it inspire you? Is your day to day activity aligned with the movement towards this vision? Does the vision align with your personal goals? Vision statements have an important place at the top of the business goals hierarchy. If they are successful they serve as a directional flag for everything the business does and they help guide everyone within the organisation. They become a powerful part of corporate culture. However if they are mere adornments they are likely to be more damaging than is imagined. Followers need to be able to respect their leadership. Vision statements that are seen as hollow, incomprehensible or directed solely towards satisfying analyst and shareholders will seriously damage respect and trust. How can followers be expected to either work toward a goal that they do not understand or respect leaders who say one thing and then do another?

The best position occurs when the corporate vision is a reality that is understood and lived by everyone as part of the organisation's culture. Sometimes we are not so fortunate as to have the powerful focus that a good corporate vision provides. However perfection is an objective seldom achieved. We have to learn to shape our own environment with the resources we have at our disposal rather than bemoaning the fact that those above us have not played their part as well as they ought to have.

Leaders at all levels should develop a vision for their area of responsibility, however small they may be. By establishing a vision you will be effectively providing the awning under which all your followers work should be performed. The objectives set beneath the awning of the vision should complement it, just as it should complement the corporate vision.

Once the vision or department objective is established it must be communicated and shared with every staff member, not just once but over and over again until everyone knows and understands it both consciously and subconsciously. To achieve this level of penetration it must be short, clear and unambiguous. Whether your message is getting through or not can be checked by regularly asking staff if they know what the department or company objective is and what it means to them. If a single person says that they are not aware of it, or does not understand it, you have a communications failure to work on.

Setting objectives

Objectives can be large or small, public or private. However if they are to be meaningful they must be specific, measurable, achievable, realistic and time-bound i.e. SMART.

Specific	clearly identified and communicated
Measurable	progress can be objectively verified
Achievable	within reach, given available resources
Realistic	is appropriate to the overall objective
Timebound	has a target time/date for completion

Figure 18 SMART objectives

Non specific objectives can be argued about forever, they lack real meaning and they are open to differing interpretations. They create noise which wastes time and effort and they should be avoided whenever possible (which is most of the time!). Non achievable and unrealistic objectives have no purpose, they too should be avoided. Measuring and imposing time constraints on objectives allows them to be tracked and real achievements acknowledged.

Imagine a shipmaster who sets the objective for a new chief officer of inspecting all the ships double bottom tanks.

The objective is not specific enough; the chief officer should know, or can easily determine, how many tanks there are but he doesn't know exactly what the captain wants him to look for. Perhaps the coatings have a history of failure or perhaps the tanks with hopper sides have experienced cracking before. A more specific objective would have been "to inspect all the DB tanks to see if the coatings are still intact"

It is measurable; they can be checked off as they are inspected with notes on the degree of coating failure being recorded as appropriate. It is achievable; the chief officer has at his disposal all the appropriate staff and safety equipment.

It is not timebound; the captain needs to define and the chief officer needs to agree on a realistic target date for completion. Without an agreed time frame and a clear understanding of the purpose of the inspection the objective is unrealistic.

SMART objectives reduce organisational noise, conflict and misunderstandings and allow people to work more effectively and efficiently.

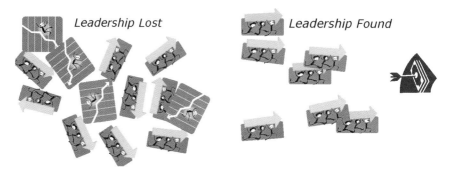

Figure 19 Finding leadership

Without SMART objectives that are agreed and communicated to those that work for you, your leadership is lost. Once the use of SMART objectives becomes an accepted practice within any organisation the power of the people is harnessed and exceptional performance becomes possible. The concept of SMART objectives is very simple, it does not take long to implement and yet its benefits in terms of time, relationships and achievement can be very significant.

Aligning personal and business objectives

To be able to work in an environment where work and personal objectives produce mutually beneficial outcomes is the ideal to strive for. It is an ideal that you can help create for those who work for you. People do things well because they want to and when they see that they are necessary to achieve their own objective. By getting this key alignment correct (see figure 19) you enhance the self motivation for the majority and performance will benefit.

It is useful to reflect on your own position and to consider whether you are in the habit of identifying and achieving your own personal goals. At the same time consider whether you regularly establish and communicate your business goals to those who work for you. Many in leadership positions fail to appreciate the importance of goal setting. It is not the same as simply telling someone what to do, in a short command. It requires mutual purpose, consideration and preparation. The best leaders are very explicit about the goals they set for their followers and their followers fully understand what is expected of them.

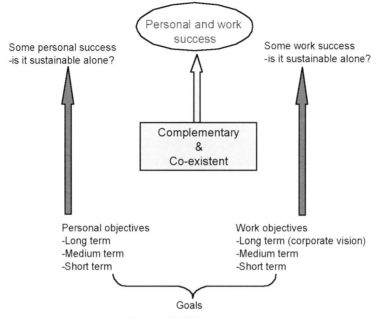

Figure 20 Effective goals

Goals can be used for close and distant things. The organisation's vision is a high level and ultimate goal. The goal to get the double bottom tank inspection completed safely and on time is a closer objective, but equally essential. The whole concept of setting objectives is not to make you and others achieve the impossible, it is to highlight where you are going and to concentrate the mind towards finding ways of getting there. Stretched objectives are important, they result in more creative tension and this drives performance. However objectives that are so stretched that they are never met lead to de-motivation and reduced performance. Objectives are necessary for rewarding business and personal lives; we all need something to aim for. Figure 20 shows how ultimate success for the individual can only be achieved when he is able to align his private objective with the vision of the organisation he is working for. When the alignment is present work assumes a meaningful purpose and performance increases as the individual is able to rise to the higher levels of Maslow's Hierarchy of Needs (figure 17, page 77). In creating the environment in which this is able to happen leaders are laying the foundations for sustainable success.

> *Objectives*
> *Whether they are for yourself or others they should be SMART stretching and agreed*

Let people know what you want their objectives to be. As objectives are achieved set new ones, keep the momentum up. Objectives provide a focus and help to keep everyone on the right path.

Values

Once you know where you are and where you want to go you need a map and some rules to guide you there. The rules are the organisation's values. They describe the boundaries of the path to success. It is useful to let the rules be 'doing words or phrases'. Staff will then always find it easy and clear to cross-reference their own performance and that of their work group against them. Two of the major objectives when establishing the path to success must be the clarity of the vision and values, at both corporate and departmental levels.

If a company's values are to be remembered and used as a point of reference for all staff and to be seen as an indication of the business culture by customers they must be concise, meaningful and clearly compatible with the vision.

Here are some examples of possible values:

- Always creating mutual trust & respect
- Swiftly responding to circumstances, customer focused and driven by constant change
- Being entrepreneurial, independent and operating with the utmost integrity

At a departmental level additional values may be things such as:-

- Constant two way communication with active listening
- Honesty and openness
- Timeliness in all things
- A "no blame" culture

The above are only examples, each operation has the potential to develop its own and then live within the territory that they define. In this way noise is further reduced and more energy is available for focused movement towards objectives. Establishing these values at team level is an opportunity for the building of mutual trust, respect and understanding.

Travelling the path

Using the vision as your target and the values as the boundaries of the path along which you wish to travel you are left with the third part of this small jigsaw and that is planning how to travel from where you are today to your vision. Perhaps this is what is called the mission. I find mission statements to be an enigma. To me they are usually, mysterious, puzzling or ambiguous. As such and with a desire to avoid unnecessary confusion on our journey I intend avoiding the use of this phrase – I think it is an unnecessary complication which I will leave others to argue about.

Travelling the path towards a vision should be the day to day or month by month activity of any business. Leadership at all levels helps to ensure more rapid movement towards the goal, whether it is an objective or a vision. In many respects this is effective strategy. For closer objectives such as small projects and day to day activities we plan our path in a much more tactical fashion manoeuvring around obstacles as they seem to block our path. Nevertheless we must still plan if we want to be successful. Not to plan is planning to fail.

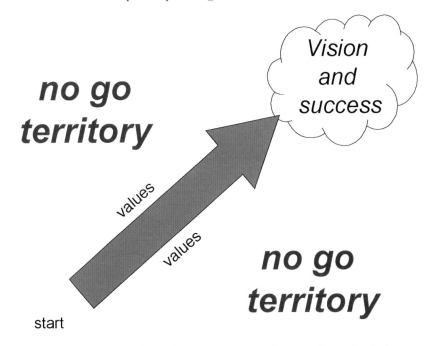

Figure 21 Vision and values, the map to success but not the end of a journey

The corporate vision and values define what should and should not be done. They should link closely to any department vision and values that are subsequently established. A dysfunctional vision statement under a corporate statement will as sure as eggs are eggs lead to a dysfunctional organisation and the road to success will be blocked. The process of working with your colleagues to establish or re-launch a vision for your area of activity and then working together to understand and clarify the associated values will go a long way to producing the focus and direction necessary for future success.

When you have agreement on what your vision and values should be, issue a detailed explanation of what you mean by them. If they are not clear you will have problems and they will soon be ignored.

Communicating the vision

Communicating the vision and the values to every individual in the organisation is the job of leaders. The best leaders ensure that this is done on a face to face basis by managers and supervisors throughout the organisation, over and over again. New staff should be introduced to them as part of their formal induction process and existing staff should be reminded of them constantly. The latter can be achieved through their use during meetings and briefings as well as on internal communications and newsletters, through posters and even through computer screen savers and wallpaper. Visions and values need to become part of the language of the organisation or department. In this way they become second nature to those involved.

It must be accepted that these documents are pivotal to everyone's understanding of the purpose of the business and its direction. They are indispensable in large organisations and without them sustainable success will be unlikely. By using such simple messages there is little room for confusion and everyone knows their own position and from there they can start working on a clear way forward.

Why have values?

Earlier we defined the values as the boundaries along the road to success. Within any organisation, if there are no values there will be no sustainable future. People working in an organisation that does not have established and communicated values will determine their own and these are unlikely to be common across the organisation. Noise will once again become a dominating feature and peoples' energy will be wasted as they operate solely within their own individual value systems. Ignoring the significance of values can be perilous whereas if they are used with absolute determination they will become the basis for a successful future.

Promoting a "no blame" culture is an important value for any progressive organisation, particularly one which needs to operate safely. In the marine industry, safety takes a high priority. However, if a blame culture exists incremental small errors and omissions will often not be reported. The consequence is that this can lead to major incidents occurring when they could otherwise have been avoided. Within shipping companies it is therefore essential that all personnel are able to speak up without fear.

It is necessary to challenge the ignorance of the leader who believes that loose federations of work groups or individuals operating to their own rules or values can bring combined success (figure 19, page 86). One ship can be well led but others in the fleet may be poorly led, perhaps giving an overall impression of a mediocre fleet.

Values and time management

The link that exists between time management and the understanding of our values and goals is of paramount importance. In order to effectively manage available time there is a need to set specific goals for us and our teams to work towards. The simplest way of tracking short term objectives is through a task or to do a list. In this basic form it is a very blunt tool usually consisting of a list of urgent, rather than, important or vital items.

Many people have written before of the need to be able to separate the urgent from the important. All too often our lives are controlled by the urgent, those things that must be done now. As leaders our example is followed. If we display ourselves as slaves to the urgent at the cost of other, more important, tasks we can expect our followers to develop the same priorities. It is a fact that all urgent things are not important, they are simply the things we are under one sort of pressure or another to rush and complete, now. The easiest example is a telephone call or a text message. More often than not they are not both important and urgent yet how often does everything else stop while we deal with them. Perhaps we like doing the task, so we give it priority, perhaps it is easy, so we give it priority, perhaps by doing it we can avoid doing something less pleasant. Working in this way is a mug's game. We are becoming ruled by being busy rather than by being effective, we are working like a donkey, not a horse. This is not a display of leadership.

In order to be able to reconsider our task lists and goals we need to return to our basic values, the things around which everything in our lives should revolve. As individuals we each have our own values, those principles and things closest to our hearts. They may include, for example, the need for financial independence or the desire to always be honest or to help others or to ensure that our children want for nothing. Values are very personal but in order to establish personal goals and meaningful business goals we need to understand theses values so that we can ensure that all the big things we do are pointing towards them. In this way we are using our time effectively. This increases satisfaction, improves self esteem and leads to improved productivity, both for you and for those that follow and observe you. Your actions, just like your moods, are contagious.

> ### The hour glass
> *Time cannot be slowed, stopped, repeated or bought –*
> *don't waste a minute – take action now*

Deciding what your personal values are is an important task and one that takes time and effort. It is something that needs to be reviewed and adjusted over time. Once finalized these values represent your ruling principles and within the limits they define, goals and daily tasks can be established and aligned. It is then necessary to determine the relative importance of each of these values to you. This can be a daunting task but it is a very important one. It is rather like Maslow's hierarchy of needs, only in this case the higher needs have to be satisfied before the lower needs. We can spend much of our time dealing with issues that relate to values that are low on our list. As a consequence we end up squandering time that should be spent focused on those things that relate to our most important values. If we attend to our highest values everything else will fall in place. If we waste too much time on lower values we will be less effective and certainly less content.

Take 20 minutes and reflect upon what could be your ruling values. Those already listed on page 88 may help you get started. Start listing them as bullet points or short phrases and follow each with an explanation of what you mean by it.

Over the next couple of days keep returning to these ruling values and adjust them as necessary. You will find that they begin to crystalise and paint a powerful picture of what your life at work and at home is really all about. Over the coming months you may wish to refine them further.

The next task is to determine the relative importance of each of these values to you. This can be a daunting task but it is a very important one. It is important for us to determine the order of our own values and then to use them as the basis for prioritizing what we do.

Our priorities should reflect our values and those that have the greatest impact on them should be at the top of our list. This approach is as applicable at work as much as it is at home.

Chapter 10

Openness: Two way communications

Summary

This chapter introduces the third finger of the Hand of Leadership. It considers how a leader communicates and how others communicate to him. It starts by discussing the importance of listening and then moves on to consider the technique of active listening. Next, the danger caused by the distractions that arise when communications are inadequate is explained. The need for proper preparation in advance of significant discussions is raised as is the need for consistency of approach and the development of compelling arguments. Trust, mutual respect, integrity and transparency are all considered in the context of openness and effective two way communications. The use of dialogue flags, the concepts of "The Daring Dialogue" and "Fight, Flight or Communicate" are introduced. The chapter concludes with discussions on handling formal and informal feedback and appraisals.

The Hand of Leadership

Awareness
of self
of others
→ **Tuning behaviour and using empathy**

Direction
defining the route
→ **Setting objectives for self and others**

Openness
sharing the messages
→ **Enabling two way communications**

Atmosphere
reducing de-motivators

Doing
Self
others

Listening as a skill and a necessity

Often discussions about communications focus entirely on speaking and yet the ability to listen to what is said is a major obstacle in communicating with others that we have all faced from time to time. A potential leader who fails to listen to others is lost before he starts. Fernando Flores, a former finance minister of Chile who, after a period as a political prisoner, obtained his PhD on speech theory, stresses that listening is not the same as hearing. Listening is an intense brain based activity that involves interrogating internally what is heard. **Hear**ing on the other hand is the basic act of noise entering our heads through our **ear**s. As an example you are currently reading this page, perhaps sitting at a desk or in a relaxed environment. All around you there is noise, perhaps it is the cars driving by on the nearby road, cargo work on deck or the gentle drumming of the main engine or air conditioning. You h**ear** these noises but do not interrogate them – you are not listening to them. But when the fire alarm goes off things are different. At first you hear the noise. Its persistence and difference causes you to interrogate it and you move from hearing to listening as you engage your brain. By this point you will have stopped reading and be, perhaps temporarily, focused totally on understanding what the significance of the alarm is; you are listening to the alarm.

If we truly listen to our people then we put ourselves into a win/win situation. They appreciate that we are listening and we learn from their dialogue. Turn things around and we have a lose/lose situation. We h**ear** what is being said but we fail to interrogate it though listening, we learn nothing while the other person is demotivated by our apparent disinterest. This is lousy leadership.

Active listening

When we engage our brain and pay full attention to the other person we are said to be actively listening. Unless we actively listen to people when they talk to us we will not gain much from the encounter. If we merely h**ear** what they are saying we are doing several things which negatively impact our function as a leader. These include, for example:-

(i) displaying a lack of interest towards the other person

(ii) failing to show respect towards them

(iii) missing any valuable knowledge and information that they can offer

(iv) wasting our time (and the other person's)

The consequences of such action can be significant. People stop approaching you and thus the amount of useful information that you glean from others reduces. The reciprocity of trust and respect decreases. Your followers begin to view you as "the boss" rather than "their leader". Their self motivation declines. The synergy of group interactions is significantly damaged. Leadership is lost.

Listen!
A good communicator is always a good listener – that is why we have two ears and one mouth!

We speak at between 100 and 120 words per minute and yet our brains can interrogate words at a rate of up to about 800 words per minute. When listening to others we have spare intellectual capacity and we have the choice of how we use it. We can day dream and perhaps plan our social life, our next leave or we can apply our brain to actively listen to the other person. If we adopted the former our listening quickly lapses to h**ear**ing with all the negative consequences that we have just seen. If we adopt the latter approach then we have the opportunity to get the most out of the intervention. We gain knowledge and information as well as showing respect and interest towards the other person. We will be helping to create an atmosphere in which others become self motivated and the potential for synergy will be established.

There are a number of techniques that can be used to improve active listening. Having important dialogues without distractions and removing many of the temptations that can cause your attention to drift away significantly increases the chances that the dialogue will be mutually beneficial. Turning your lap top or PC off or moving it out of sight helps to retain your focus, a new incoming message can cause your attention to divert. Mobile phones should be off or turned to silent mode. Do not answer them during dialogue sessions. Consider where you have your dialogues and avoid interruptions from other parties. Show the person that you are interested by establishing good eye contact and using positive body language. Offer some words of encouragement and paraphrase parts of their message back to them so that they can see that you understand and are really listening to what they are saying.

Active listening, like any skill requires persistence and effort. Practice these simple techniques and the respect and free flow of information will increase. Mutual understanding and knowledge will be shared between you and your colleagues. It seems silly not to adopt such simple practices when they can really make a difference.

If you remain sceptical about the importance of active listening look back over the last two paragraphs from the perspective of you as the person doing the talking and your boss being in the role of the listener. Ask yourself two questions.

(i) How would you feel if the listener appeared more interested in reading the report on his desk or dealing with interruptions than in actively listening to you?

(ii) Would you feel better if the listener was using the active listening techniques described?

Questioning

The way we pose questions defines much of the way in which people respond. For example, an aggressive demand is likely to be met with either an aggressive retort or stunned silence. The middle ground of effective communications will almost certainly be lost. Similarly the type of questions we use establishes the way in which the dialogue progresses.

Closed questions are those that can be adequately answered with a "yes" or a "no". They do not generate much meaningful information, other than agreement or disagreement and they do not encourage participation in the dialogue. On the other hand open questions require an explanation or an opinion to be given and this provides the listener with a much richer source of information and generates participation. Leading questions are those that lead the speaker into a response without it being an opinion or belief which they necessarily hold. For example "I believe that this meeting is a good idea, don't you agree?" The questioner is expecting the listener to agree with them.

Interrupting the other person is another technique that should be used sparingly. You do not know what they are going to say next and by stopping them you may create a feeling that you are not really interested in the conversation. However, there are occasions when dialogues have to be re-focused or re-directed and on these occasions interrupting may be acceptable.

Leadership as a two way dialogue

The use of active listening techniques demonstrates a willingness to include others in your decision making processes and your desire to respect their contribution as part of your work group. It has already been established that leadership is not a solitary activity; without followers a leader cannot exist, it really is a case of "it takes two to tango!"

The whole concept of leadership is about much more than just the leader. As Jonathan Gosling, Director of the Centre for Leadership Studies at Exeter University, says

Where L are leader(s) and F are follower(s)

$$L \times F = leadership$$

but if L or F is zero there can be no leadership

As a leader, at what ever level in your organization or ship, there is significant benefit to be gained by showing interest towards your followers and there is no better way to do this than to establish a regular dialogue with them and to use the techniques of active listening. Their respect for you will increase and their own self esteem will grow. You are giving them the opportunity of helping you to succeed. Never forget, the people have the power.

Communicating objectives

It is the CEO's job to get the vision to the people. However it is also the responsibility of the leader of any grouping within the organization to establish the goals (vision and objectives) for his area of responsibility, however small his area of responsibility may be. These goals or objectives need to align broadly thereby forming an important part of that larger jigsaw.

Leaders have the task of establishing the objectives for their followers. In order to do this many see the value of working with their followers to define the goals in

> **Communicate with confidence**
> "As soon as you move one step from the bottom rung of the ladder, your effectiveness depends on your ability to reach others through the spoken and written word" Peter Drucker

the first place, thereby achieving two benefits. First they develop objectives that are aligned to the expectation and desire of the followers and secondly this alignment brings a higher level of commitment from the followers towards their eventual achievement.

It is the leader's responsibility to communicate and explain the goal(s) to his followers. If he fails to effectively explain why there is a need to head in that particular direction he should not be surprised if the outcome is unsuccessful. Simply writing a memo or sending an email announcing the goal(s) is seldom adequate. Yet, many leaders seem to think that indirect and singular communications are adequate. In their minds everyone should understand what they have said and remember it without any further repetition or explanation. Such leaders often make the erroneous assumption that their followers are bound to think in the same way and that their single communication is explicit enough to clearly convey exactly what is required. Look at the statement and below and then answer the questions that follow it.

A man walked into the garage through a narrow door, he was wearing a suit. The garage owner then turned on the electricity. Shortly after that a compressor started and the lights came on. A little later the garage owner was able to pump up the car's tyre. Suddenly a person, in a blue overall, jumped into the car and raced away, just missing the garage door. The owner called the police.

1. The garage owner was a man   y/n
2. The compressor was electrically driven   y/n
3. Someone turned on the electricity   y/n
4. The thief stole the car and drove it out of the garage, just missing the door   y/n
5. The car nearly hit the garage door because it was being driven fast and the doorway was narrow y/n
6. The police will try to find and return the car to the garage owner   y/n
7. The garage owner called the police   y/n

Now look at the answers in appendix 1, do you agree?

All messages can be ambiguous if viewed in isolation.

Look at these examples and try to understand them;

1. *Pack Ice* – is this a description of a type of ice found in arctic regions or is it an instruction to Mr. or Ms. Ice?

LEADERSHIP 97

2. *Red tape holds up new bridges* – does the tape support the bridge or do they mean bureaucratic procedures?

3. *Giant waves down Queen Mary's funnel* – was the funnel knocked over or did water go down it?

4. *New study of obesity looks for larger test group* – were they looking for larger people or more people?

5. *The Duchess handled the launching beautifully, confidently smashing the champagne against the prow. The crowd cheered as she majestically slid down the greasy runway into the sea* – who or what slid down the runway?

We fill the gaps that we believe to exist by making assumptions in an attempt to satisfy ourselves that the message is not ambiguous. In reality we tell ourselves a story and everybody's story will be different, the ambiguity is increased. Given a different story everyone will draw their own, different conclusions. If leaders throughout organisations fail to clarify goals and messages to their followers it would seem unreasonable for them to expect these followers to work effectively towards them. We are moving towards the chaos of poor communications.

Important issues such as goals and values need to be clearly articulated and established in the minds of everyone involved. This is a major leadership task which if done successfully harnesses the power of the people very effectively. They know were they have to go, they know why and they remember this from day to day. Our brains are not faultless and we find it easy to forget issues and facts unless we are able to convince ourselves of their importance, either consciously or sub-consciously. In communicating the objectives leaders need to fix the messages into the minds of their followers.

Clear facts with clear explanations are the most memorable, so keep them simple. Remind people over and over again what the goals are and why they are so important. Repetition is an essential ingredient in a successful leader's repertoire. Even if you think you have communicated the goal(s) enough, you have not; keep talking about them, engage your followers with communications that are effective. Listen to them and discuss their thoughts. Engagement brings ownership and commitment.

Constant communications

"We communicate in everything we do, we communicate in silence, we cannot not communicate."

The next time that you are talking, formally or informally, to a group of people study them. Look at their expressions, look at the amount of eye contact they have with you, look at their body language, look at what they are doing. Without exception they will all be communicating to you, whether they realise it or not. Ask an awkward question of an audience and see how many of them try to avoid eye contact with you – they are communicating to you that they do not want to be the one to answer – they have gone into hiding. Someone else in the audience falls asleep, his message is "you are boring, I am turning off".

An audience or an individual that is not engaged with the speaker will have difficulty maintaining an active listening style for long. It is your job to keep those you talk to interested, you must keep them engaged. Keep your message relevant and simple. Repeat it, speak slowly and clearly.

All effective leaders are communicators who constantly explain why followers need to do things. At the same time they listen and involve them in their decision making processes. They always show interest and are aware of the need to offer praise when things are done well. Supervisors, managers and officers no longer need to be aloof from their followers. The days of 24/7 command and control and autocratic leadership have been assigned to the bottom of the ocean.

Being consistent

It has been proposed that leaders need to be trusted if they are to be sustainably effective. Those that show a mismatch between what they say, what they do and what they demand of others will not be trusted for long. Their approach is full of inconsistencies. Over time uncertainty, confusion, blame and perhaps fear (of doing the wrong thing) will dominate the workforce. Trust will evaporate and followers will operate in a self protecting mode, focusing on survival rather than excellence.

Consistency builds peoples confidence and this in turn creates trust - that indispensable feature of a leader. People can understand the way you operate, albeit with surprises from time to time! It is no good expecting others to trust you if you demonstrate an inconsistent approach. By being consistent in your actions your message will be clear and its value reinforced.

If you operate in a consistent fashion others will feel able to trust you, even if they do not always agree with what you are doing or saying. They come to understand how you will react to issues and challenges and with the benefit of this knowledge they are more likely to become team members who feel comfortable challenging your thoughts and actions.

If followers feel that the leader cannot be trusted the success of the leader will be limited. Why? Because in the absence of trust there will be an atmosphere of mistrust and followers will waste time and effort manoeuvring to ensure that they do not fall victim of the leaders inconsistencies.

Imagine that your boss tells you that he intends giving you a good pay rise at the end of the year as he has noticed that you have been working very hard on a new project and that your salary package is currently a little low. The end of the year review arrives. You receive a very good appraisal but no salary increase. How would you feel towards your boss?

Taking things a step further you decide to ask for a meeting to understand why you have not received the increment. At the meeting you respectfully confront the boss and remind him of his earlier promise. He offers an apology and tells you that though you did a great job he can not give you a raise because he simply does not have the budget. He goes on to say that anyway, he did not really mean to give you the impression that he would definitely give you a rise at year end, he felt sure that he had only said he may be able to give you the rise; there was no promise. How do you imagine that you are feeling now? How will you view this boss in the future and what will you think of any future promises he makes?

Another example – you are scheduled to attend a meeting with a group of your colleagues at 1400 on Tuesday. Your manager is going to chair the meeting. You recall that at the last meeting several of your colleagues arrived late. The manager was annoyed and demanded that you all arrive on time for future meetings so that his valuable time was not wasted. You were not entirely happy with the way the manager had handled the situation but you agreed with the need for meetings to start on time. The reason why you were never late for meetings yourself was because, you showed respect for others that were going to attend and avoided wasting anyone's time (not just that of your boss). The meeting time arrives and you are pleased to see that all your colleagues are there on time. However your manager is not present. Fifteen minutes after the prescribed start time you decide to call him. He answers the call and declares that he is busy and will be at least another fifteen minutes. He eventually arrives thirty minutes later and starts the meeting without an apology.

Your manager's message to you and your colleagues has been inconsistent. On the one hand you have been instructed to be on time for meetings while on the other he has demonstrated his ability to ignore his own instruction. This is not leading by example nor is it inspirational for followers. Ask yourself how you would feel in these circumstances. His approach may annoy you and some of your colleagues sufficiently to devalue the meeting significantly. How about the level of respect you now feel towards your manager?

A leader needs to be consistent in all he does, the way he talks to people, the way he communicates with them and in the way he acts. If he shows favouritism to an individual or group it will be noticed by those outside of the "in group" and they will quickly lose trust in him because of his inconsistency of approach. They know that his 'favourites' will always receive his special attention.

An interesting consistency in leaders is what I like to call an ability to be "consistently inconsistent". This occurs when followers know, without having to be told (therefore it is consistent), that the leader has a tendency to do certain things at times and places that they cannot always predict (therefore it is inconsistent). Leaders do not need to announce their every move, they need to avoid being creatures of habit. A degree of surprise in where they appear and the things they do helps to keep people on their toes. It is perhaps unusual for a watch keeping chief officer or second engineer to appear on deck or in the engine room in the middle of the afternoon but the occasional such visit has the potential of (i) showing their staff that they are interested in how things are going even though they are 'off duty' and (ii) it creates a little tension that serves to keep people focused. Do it too often and you may be seen as not being prepared to trust the team to do the job, do it too seldom and you may be allowing the unexpected to occur. Get the frequency just right and the people remain on their toes and respect you because of the interest you are showing in them.

Trust and mutual respect

Perhaps the greatest endowment that a leader can give any group of followers is a culture of mutual trust and respect. No one can lead effectively unless they are trusted. This applies equally to the most senior people in organisations as it does to line managers and supervisors. Look at Paul Wolfowitz head of the World Bank, until May 2007 when he resigned, "in the best interests of the bank". He had lost the trust and respect of many of his staff through the allegedly unfair and unapproved salary increase that he had given to his long term partner who also worked within the bank. His leadership floundered when the perception of others led them to stop trusting him. It is important to remind ourselves again of the power of perception; it is the people who ratify someone as their leader. Whether he had ever done anything that was technically a significant contravention of bank rules we will never know. What we do know is that his actions caused many of his followers to denounce him as their leader, and he paid the price. The people have the power.

The people must trust the leaders and the leaders must trust the people. Once this culture has been established significant energy can be focused on the push towards the common vision. Consider how much disruption there was to the normal business of the World Bank during the weeks running up to his resignation. Thousands and perhaps even millions of dollars worth of time must have been wasted on chatter within the halls of the organisation and in the offices of its governmental stakeholders. If mutual trust is absent the full potential of a work group can never be achieved and leadership is lost. Similar examples can be found at every level within organisations albeit that they do not all make the world headlines. Trust is easy to lose but it is much harder to regain. Once lost it is often gone forever.

Seafarers, of all ranks, who operate today under the command and control regime of previous centuries are unlikely to be full of trust and respect for their leadership. Consciously or subconsciously, openly or subversively they will be wasting energy by creating noise. In earlier chapters we discussed the negative impact that bullying can have on people. They withdraw, they internalise the problem and seek revenge in one form or another; people do not trust bullies and

---

Command and control
I am in charge, you do what I tell you, no questions.
What value does this approach have today?

autocrats. Trying to lead mature adults with a command and control approach in a non military, non emergency situation today is a sure recipe for failure.

Mutual trust will remain absent in organisational cultures unless all parties within the organisation fully appreciate that integrity it is not an 'optional extra'. We unfortunately live in a world in which integrity is not adequately valued. There are hundreds, if not thousands of examples where integrity has been cast aside in favour of short-term profit or personal gain.

If people trust you in your workplace, they will respect you. You will have earned their recognition because you offer them similar recognition. Trust and respect are mutually created and shared traits which either party can irrevocably harm through any act of insincerity. Protect the respect you have for others delicately, it is easy to damage its mutuality.

Integrity

Trust and respect flourish when integrity is high. If integrity is absent or optional, trust and respect will evaporate. Integrity, in the context of people, is the act of behaving in a way which complies with moral and ethical standards. People who behave with integrity do not do immoral or unethical things. The practice years ago of discharging oily slops at sea during the hours of darkness to avoid detection illustrates low integrity from the officers involved. But even more significantly it questions the integrity of ship operators and owners who choose to turn a blind eye to the procedure rather than enduring the delays and costs associated with slower tank cleaning processes. An appraisal of personal integrity can be difficult. It is very easy to insist that we are of the highest integrity whereas in reality we may be in denial. Ask yourselves the questions below and consider your responses carefully and be honest with yourself;

(i) You are having some repairs done on your house and the contractor says that if you pay cash he will not charge you the sales tax. Would you

   (a) pay in cash and save the tax

   (b) tell the contractor that you are not prepared to pay cash and accept the larger bill?

(ii) A supervisor comes to you and says that he has found an error on an invoice to a minor customer and that there was an overcharge of 15%, the customer has paid and the file has been closed. Would you

   (a) tell the supervisor to ignore the error

   (b) instruct the supervisor to contact the customer and arrange a refund?

(iii) You are taking part in a sailing race and graze a mark of the course as you round it. This is a rule infringement that requires you to re-round the mark before continuing. No one has seen the incident. Re-rounding the mark would mean you would drop three places in the race and miss the chance of being a prize winner. Would you

(a) carry on and tell no one about the incident

(b) re-round the mark?

(iv) With some friends you enjoy a meal out. It is your turn to drive home but you realise that you have had a little more to drink than you should have and you know you are just over the drink drive limit. Would you

(a) drive home

(b) call a taxi?

(v) Your wife phones you at work to remind you that you have to write a letter to your daughter's school that evening. She reminds you that there is no printer paper left at home. The stationary shop is a 5 mile drive from the office in the opposite direction to your house. Do you

(a) take a few sheets from the office printer

(b) drive to the shop and purchase a supply of printer paper?

These questions and your individual answers are not the issue here; you know what you would do in each case. The purpose of the exercise is to provoke some reflection on the issue of integrity. Now look at the same scenarios but consider the repercussions if option (b) was taken on each occasion;

(i) The contractor gets prosecuted for tax evasion and you get called as a witness.

You have already lied.

(ii) The next day the customer approaches the supervisor and points out the overcharge.

One or more people will probably lie as this situation is now addressed.

(iii) You finish the race first but then are subject of a protest because your mark hitting had actually been recorded on video tape.

You have already lied.

(iv) You have a minor accident and fail a breath test.

It is likely that you will now lie and claim that you did not realise you were over the limit.

(v) Your manager catches you as you surreptitiously put the printer paper into your briefcase.

You will probably tell a lie.

In each of the above instances you have jeopardised or lost the trust of an individual or group of people and your future actions will always be considered questionable. During the process of extracting yourself from the situation it is quite probable that lies will be told. As Aristotle once said:

> "Liars when they speak the truth are not believed"

If you are not believed you cannot lead.

The purpose of this exercise is not to attempt to determine your level of integrity, only you really know that. It is designed to illustrate the importance of trust, integrity and honesty in any position of responsibility. Integrity is often claimed as a value of a company, organisation or individual but too often it is just window dressing, writing and saying what people want to see rather than living the value.

> **Do not live the lie**
> Integrity can be defined as "adherence to moral principles" whereas failure to operate with integrity could be defined as "having to live a lie."
> Good leaders do not live a lie.

As individuals we have choices to make in our journey through life and finding our position in respect of integrity and trust is one of these. Leaders need high levels of integrity. Most leadership gurus identify integrity and trust as two of the most important elements of leadership. They are without doubt the most important values that a leader can have.

Building trust is not an overnight activity; it takes time, commitment and determination. When a new leader arrives in the workplace there is a natural feeling of distrust towards him. This new leader may, for example, be a supervisor of a small team where the previous leader has moved to a new position. The team

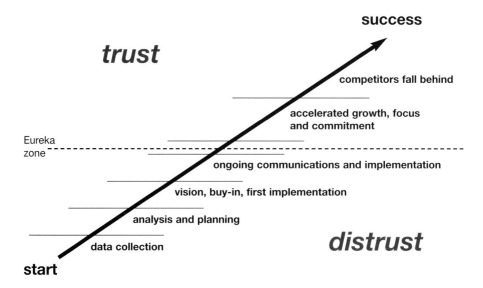

Figure 22 The Eureka zone

are anticipating the arrival of the new supervisor and they are feeling uncomfortable. They had developed a good relationship with the previous incumbent and did not want him to be moved on. The prospect of the unknown is always unsettling and new leaders have to tread carefully in their early days to build support and bring their followers onto their side.

The new leader's task is to quickly collect and analyse the data he needs to establish a vision and secure the buy in of his followers. He has to build relationships and gain respect and trust. Only then can he move through the eureka zone (figure 22). This occurs when the actions and style of the new leader is understood by the majority and they have bought into the vision and objectives. The new leader's initial target must be to bring his people to the eureka zone as quickly as possible. Once reached the business potential can be realised and the competition left standing.

Transparency and clarity

The flow of information through any work group is the glue that binds the parts together. We talk elsewhere about communications but in this section we will reflect on the need to share knowledge with the widest audience in your work environment.

To lead we have already established that you need to have the trust of those you work with. Generating trust requires transparency. In turn being transparent demonstrates your trust in others. It also instils in others a feeling of being part of the bigger picture. But there is more than this need. People must have information to manage and grow their business effectively. To impose a vision or objective on others and yet exclude them from significant data and information that supports the vision is a serious limit to potential. To hold someone accountable for financial performance of a unit or ship when they are not given the full detailed management accounts is like asking an engineer to fix the generator without visiting the engine room. He knows there is a problem but does not have access to enough information to draw a meaningful conclusion that will enable him to rectify the situation.

People may fail to share information for a number of reasons. It may be because they;

(i) believe knowledge is power; the dictating leader

(ii) communicate with you on a need to know basis only: the secretive leader

(iii) believe you will not understand: the arrogant leader

(iv) are afraid you will tell the competition: the distrusting leader

We have already suggested that the above leadership types are negative. They will never produce long term sustainable performance.

There are, nevertheless, some circumstances in which you have to be selective with information in order to protect your true business advantage, but these cases are rare. Most of the time freedom of information must be a matter of course and routine. The downside of the occasional loss of confidential data must be carefully weighed against the failure to share data in the first place. Trust soon evaporates if you keep people in the dark.

Any organisation should be professional as should the staff it employs. The organisation needs to be made up of people who are able to sensibly handle and manage the information that they are provided with. If within your own work group the feeling is that there are key staff members who cannot be trusted with anything other than the absolute minimum of information there is a serious problem. The problem may however have two sides and there is a need to be frank when considering this issue. Simply put, either you are too secretive and distrusting of your staff or your staff really cannot be trusted. Do not deceive yourself that the problem always rests with the staff. Perhaps you have a tendency to keep people in the dark. Only you know. If you see your own weakness in this area, admit it and start changing. Your future and that of those who work with you will be better for it.

One way of strengthening your leadership is to help others to understand that they are valuable components of the business or team and that you are willing to be transparent with them. You can achieve this by demonstrating your trust in them. If, over time, they are unable to handle sensitive information in a sensitive manner you will however have to make some tough calls. Never forget that life is not meant to be easy. There will be hard decisions to make as you build and lead your team to a sustainable and rewarding future. If on the other hand things continue with most people being kept in a sterile vacuum then leadership has failed and the operational effectiveness of the unit will be compromised, it will either fail or at best under-perform.

Ask yourself some questions and reflect on your responses. Are you already a trusting and transparent leader?

(i) Do you insist on being copied on routine correspondence just in case people say something that they should not?

(ii) Do you share the budget or your targets with your direct reports in their entirety or do you only share them in a selective fashion?

(iii) Do you let your direct reports deal with most issues without reference to you or do you always insist on checking their conclusions before you let them move on to implementation?

(iv) Do your direct reports regularly pop into your office or cabin to bounce ideas off you?

(v) Do you let your direct reports know the major issues that are worrying you?

Success depends on team effort. Team effort depends on trust and trust does not exist without transparency. The eureka zone will remain over the horizon if there is limited transparency and you will never be trusted or able to mobilise the real power of people.

Daring dialogues

Sometimes we need to have difficult discussions with people. Maybe their performance has dropped or you need to talk about an assignment you know they will not be keen on performing. Perhaps you have to talk to them about their behaviour towards their own staff or maybe your own boss has planned a discussion with you to explain why you were not promoted when a recent vacancy came up. We live in a world full of opportunities for dialogue and though these opportunities are an ideal opportunity for two sides to reach a mutual objective they often end up as fierce battles and often with casualties on both sides. It has been said that the human mind is seriously flawed and whereas people can communicate with music, dance and writing when we have to talk to each other something changes and we often move towards an adversarial approach of attacking and defending, reason seems to be rejected.

These potential battle fields of words are what I like to call "Daring Dialogues" and they need careful handling. Battles and their unfortunate consequences can be avoided but it is not always easy. In a Daring Dialogue the respect you have for each other is at risk and the quality of your future relationship will be defined by the outcome of the dialogue.

When dialogue fails we move to flight or fight. This is where we focus on attacking; defending or withdrawing. Dialogue is no longer our objective. Flight and fight, a phrase first coined by the American Psychologist Walter Cannon in 1929, can be considered as being at the ends of a line that covers the breadth of our potential behaviours in dialogue. Flight occurs when direct confrontation is avoided, yet it is still confrontation. The person being attacked withdraws either by removing himself from the dialogue physically "I have had enough of this, there is no point in continuing this meeting" or he disengages from the dialogue and offers no further contribution. He uses silence as his weapon. The person in flight has chosen to remove himself from the situation and the dialogue collapses. At the other end of our imaginary line the dialogue becomes a painful verbal duel where each operates with emotions unchecked. Quickly respect for each other evaporates – the dialogue has collapsed.

Dialogue is different from debate. It is approached with an open mind, seeking to find the best way forward. The participants share a desire to reach the best mutual outcome; they share this as a mutual objective. When the common objective is lost and things turn to a need for personal victory, the opponents, for that is what they have become, move towards flight or fight and dialogue has ended in failure. In a dialogue neither side should be looking to "win". In debates this is not the case. Each person enters the process with a fixed position and with the intention to persuade the other to adopt it. Debating is duelling and people get injured in duels unless there is a moderator or referee to control excesses. Debating can be useful in helping third parties to formulate their opinions. However, the two debating parties are unlikely to get much personally from the battle, other than perhaps gaining or losing third party support. Their personal views are unlikely to change. Look how politicians gain and lose support on the basis of how they perform in a debate.

If a "Daring Dialogue" is necessary it is important to prepare yourself as much as possible in advance. First you have to plan what you are going to say and anticipate how the other person is going to react and also how you may then react to them. Your primary concern should be to enable the event to remain a dialogue. Daring Dialogues are by their very nature high risk events, there is a lot at stake and it is easy to shift from dialogue to duel if you are not careful. When there is high risk there is tension and it is easy for a Daring Dialogue to become an aggressive fight where someone or perhaps both people get harmed. If this is allowed to happen there will be no winner. Today's human beings are biologically similar to their ancestors who thousands of years ago often had to physically fight to ensure their survival. Darwin's theory of evolution was based on the presumption of the survival of the (physically) fittest. Our ancestors fought with those who challenged their security and life was often a series of duels, which were sometimes fatal.

At times of conflict, even in dialogue, adrenalin is released in our bodies and this increases the blood flow to our muscles enabling us to use physical strength to defend ourselves. An unfortunate consequence of this is a corresponding reduction in the supply of blood to our brains, at such times. Thankfully today physical action to defend our opinions is generally unacceptable, yet we retain the physiological characteristics of our ancestors and if we feel threatened adrenaline still enters our system. When this happens our brain is detuned and we begin to lose the ability to use logic and reason. If we lose control during a Daring Dialogue our ability to conduct a coherent, logical and emotion free discussion ends, our brain is beginning to shut down. We say things that we do not mean and we say things that we later wish we had not said. All in all we make a fool of ourselves. The Daring Dialogue has collapsed into a duel where people get hurt.

The Flag for Dialogue (figure 23, opposite) is used to remind us how to conduct Daring Dialogues effectively and to alert us to the consequences of failure. Such dialogues work if you are able to keep within the area of common understanding, the green zone.

When both parties are aware that the purpose of the Daring Dialogue is to discuss a sensitive matter and to reach agreement on the mutual objective, it is possible to keep the dialogue in the green zone where each respects: -

- the issue
- the needs of the other
- the mutual objective

While both remain in the green zone there is common understanding and working towards the mutual objective can continue without threat to either party.

If one party fails to respect these essential requirements they will begin moving from the green zone into the orange zone. Common understanding is now replaced by concern. This may be felt by either or both parties. The danger of movement into the orange zone is that if it goes un-noticed by the offending party he will continue to damage the dialogue and eventually someone is likely to find themselves in the red zone; the dialogue will have failed.

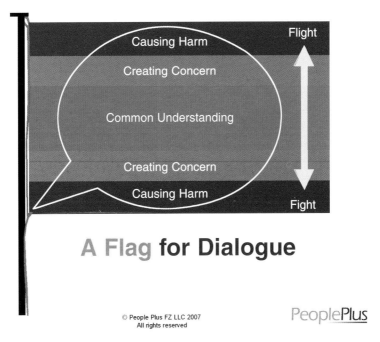

Figure 23 A flag for dialogue

Within the area of common understanding where respect remains high, trust is maintained and the fear of conflict is contained. Movement towards the mutual objective is assured.

One of the major behaviours that can derail a Daring Dialogue is defensive dialogue where one party begins to drift outside of the green zone towards the orange warning zone. This occurs when either party starts to defend his position from an emotional perspective rather than from a constructive and logical perspective. A constructive, non emotional and logical approach prevents the dialogue from becoming threatening. When one party begins to focus on personal defence, the mutual objective is lost and the quality of the dialogue deteriorates. Personal defence in dialogue may appear in two forms. The first is displayed by one person refusing to listen to the dialogue anymore. The dialogue is suspended and a monologue of reasons why you are right and the other party is wrong ensues. The other defence is to attack and things begin to fall apart as accusations come to the fore and a culture of blame is established. One side of the dialogue has now moved to the orange zone on the Flag for Dialogue. The danger signs are easy to spot in someone else but they can be more difficult to spot in your own behaviour.

Read the story below.

Cargo overloading
A Daring Dialogue or a Failed Dialogue?

John takes a seat in the dayroom and can see that the captain is stressed. He realizes that this is going to be a Daring Dialogue and he is determined to keep his emotions in check. The loading had been long, over 36 hours and since it was only a part cargo of fuel oil several of the tanks were going to be slack on completion. The charterer had instructed that they were to load the nominated quantity but they requested that the final amount should be as close as possible to the upper end of the +/- 10% tolerance i.e. the ship had to load the nomination +10%, but no more. Towards the end of the loading the captain was pressed by the charterer to finish before darkness in order to get to the next port in time to unload prior to another vessel arriving to take the berth. John worked and reworked his figures and final topping off was limited to three tanks, each would be slack (not full) on completion. The captain had stressed to John that they need to be able to sail before darkness as he had the office and the charterer on his back. With this in mind John was keeping the loading rate higher than he liked to for the topping off. When he got down to the last tank he reduced the rate and brought the tank to its final ullage and called for the loading operator to stop the loading. John was pleased, all looked good and the final tank had stopped exactly on the ullage he had wanted. When the cargo surveyor did his rounds to check the final quantity for the Bill of Lading he found one of the final three tanks to have several tonnes more in it than it should have had and consequently the Bill of Lading exceeded the +10% that was requested by the charterer and allowed by the charter party. The ship was in default of its obligation to the charterer.

No one was going to be pleased with John about this.

The ship's captain called a meeting. Imagine that you are John, his second in command, the chief officer. He has asked you to come to his cabin so that you could discuss the problem that there now was with the charterer. The loaded quantity of cargo had exceeded the +/- 10% tolerance that was included in this particular voyage charter.

Captain: "Hello John, have a seat… John, what on earth happened? I have already had head office and the charterer on the phone. The charterer is saying that the receivers may not now accept the cargo at all. Why on earth didn't you check the automatic-gauges on the tanks before you topped of. That is a pretty big mistake isn't it?"

John "It was an oversight, a combination of my tiredness and the pressure we were under to get finished before darkness. The pumpman is with the engineers right now trying to see what caused the gauge to stick. I'm very sorry."

Captain "Come on, don't make excuses. I clearly told you before loading that the 10% tolerance must not be exceeded. You have let yourself down on this one."

The captain was in no mood for a Daring Dialogue. He opened with accusations (blame) before he had been able to share any common understanding with John. As a consequence John also moved away from a common understanding of the objective and chose to focus on self defence. This dialogue had already moved to a duel and it would probably just get worse and yet at the end what would it have achieved? Nothing positive I fear. A blame and defence battle where respect for the common objective was lost the moment the dialogue started.

Let's take a re-run of the meeting and see how it might have started in a better manner.

Captain: "Come in John, take a seat…. We've got a problem now, brief me on what happened. We need to work on a response to the charterers and head office before they get too excited"

John: "I'm sorry to have put you in this position. The loading was going well but I tried too hard to get finished in time for the daylight sailing, I should have topped off at my usual rate. I've got no excuses. The pumpman and the engineers are checking all the gauges as we speak, but I know its too late now. What do you need from me for the report?"

Captain: " I can't say I am pleased with this outcome but I know you well enough to recognise that it is not representative of your usual performance, so let's just focus on the report we need to prepare and see how we can mitigate the consequences for us all."

This time they have stayed in the area of common understanding where each respected:

- the issue
- the needs of the other
- the mutual objective

Achieving this success in dialogue is not always easy. Both participants need to recognize the ease with which a dialogue can turn into a duel. In the event that one participant moves from the green zone of common understanding into the orange zone of concern there are three ways that the dialogue can now shift:

(i) Stepping back into common understanding

The participant in the orange zone realises that he is causing concern and regains control of his emotions and behaviour. Having done this he can once again respect the issue, the needs of the other and the mutual objective. This is a useful time to offer an apology. Events re-enter the area of common understanding; the Daring Dialogue is able to continue.

(ii) Being pulled back into common understanding

The green participant becomes aware that his colleague has moved into the orange zone. Nevertheless he remains focused on the mutual objective and coaxes his colleague to return to the area of common understanding.

(iii) Duelling

The orange participant remains unable to accept that he has abandoned common understanding and eventually his belligerence pulls his colleague away from common understanding. Together they rapidly move into the red zone and start to cause harm to each other. The dialogue has ended with both participants entering a flight or fight mode. There will be no winner.

Managing a Daring Dialogue:

- Prepare properly
- Choose a time and location where you can give 100% attention
- Control your emotions
- Listen to what the other person says

- Avoid being judgemental, particularly, before the meeting starts
- Remember it is a two way dialogue
- Treat the other person with respect and dignity
- Descibe – start by describing the situation as you see it
- Awareness – be aware of the other persons feelings and reaction
- Results – state the type of result you need from the meeting
- Evidence – have the evidence at hand

It is not always possible to prepare for a Daring Dialogue in advance. Sometimes we have a routine meeting and suddenly something sensitive or controversial comes up and that we find ourselves on one side of the Daring Dialogue. On these occasions a mental referral to the Flag for Dialogue helps alert us to what we need to do to stop the dialogue from ending in failure. On other occasions we know we are going to have a Daring Dialogue with someone and then we have the opportunity to prepare. Figure 24 offers a useful set of guidelines for ensuring that the dialogue has a fair chance of survival.

Giving feedback

The skills associated with the flag for dialogue can be effectively utilized in most of the encounters that we have with people. With one eye on the objective and the other on ones own behaviour and emotions it is possible to step into a dialogue environment that is both non-threatening and rewarding. The skills are particularly valuable when handling appraisals with staff and colleagues, whether they are formal or informal. These are traditionally highly charged Daring Dialogues that are prone to go wrong. They can end up as monologues with the recipient choosing the silence of flight as the best route. This approach produces only negative results. The recipient is likely to leave the meeting de-motivated and at worst with the intent of causing some sort of subversive harm to his manager. Read the example below and make sure that in the future your cup is always clean!

The third mate – a leadership failure

The ship's third mate, Ian, is on his first voyage after getting his certificate. He is feeling good though nervous about the expectations that everyone has of him; no more the luxury of the responsibility free life of a cadet. The first couple of months of the voyage had gone well, or so he thought. He was getting on well with his colleagues and the old man seemed fairly relaxed. However he had got off onto the wrong foot with the steward who looked after his cabin. He didn't think that he was doing a good enough job cleaning it, some days he just seemed to make the bunk and no more – the shower never seemed to get a good clean. After a couple of weeks he'd spoken to the steward. He thought he was firm but fair with him. The steward didn't agree and began to argue back. When this happened the third mate felt he must assert his authority and insist that the steward does a better job in the future. George the steward was a good 20 years older that Ian the third mate and he has seen it all before, a new junior officer wielding new found power and he had had enough. George turned to flight and stormed out of the cabin. Ian felt embarrassed and thought "what the heck, I'm in charge, he is just a steward". The encounter did not start as a Daring Dialogue, there was zero common understanding, it was a duel and the steward's exit to flight certainly did not mean he was the only loser.

The same evening the 3rd Mate came down early for dinner and was joined by the second mate and a couple of the engineer officers. Ian had a light meal and thought he would finish with a quick coffee before rushing to the bridge to relieve the mate. George was waiting table and went off to get Ian his coffee. Once in the pantry he poured the coffee, but before leaving the pantry he spat into the cup! George was having the last laugh and the 3rd Mate would never know. George was enjoying revenge and perhaps there would be more opportunities in the future. Moving to flight does not mean all is ok. On the contrary mutual respect has evaporated and unexpected consequences are likely to be the result for both parties.

Sometimes when one party lapses into creating concern or causing harm the other may still be able to rescue the situation without immediately pulling the other party back to common understanding. Instead he has at his disposal the option of suspending the dialogue. This is not the same as moving to flight. On this occasion the common understanding of the one party leads him to suggest a temporary suspension of the dialogue. He schedules a resumption of the Daring Dialogue at some time in the relatively near future. The intervening period allows emotions to discharge and gives the other party the opportunity to return to common understanding. The adrenaline rush is halted.

Daring Dialogues do not force us to hide from the issues. They allow us to face them in an environment that is secure and full of common understanding.

As leaders become good leaders they encourage those around them to share information and act as if they were one. In recent years, the power of effective teams and their success at delivering results has made them one of the best tools for business growth. Your ability to lead effectively within a team environment is crucial.

Chapter 11

Atmosphere: Avoid de-motivating

Summary

This chapter focuses solely on how to create an atmosphere in which other people want to excel. It introduces the fourth finger of the Hand Of leadership. The idea that all leaders need to do is motivate their followers is dispelled. The power of reducing de-motivators is emphasised as an important leadership tool and skill. Enthusiasm, passion and optimism are also discussed as ways of improving the atmosphere in any organisation. The problem of voids in leadership hierarchies is raised along with the need for the avoidance of unnecessary bureaucracy.

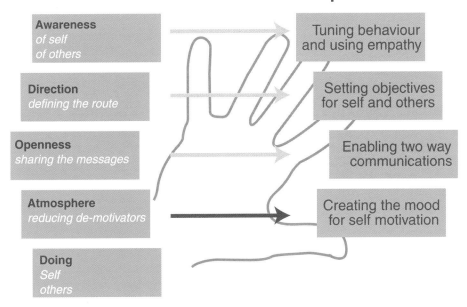

Motivation as a quick fix?

It will be hard for you to find a chapter in another leadership book that includes the word "de-motivation" in its title, though you will find plenty that include "motivation". It seems that we are operating in an imaginary leadership reality that is heavily biased towards our personal unique reality. Motivation pops up everywhere as the panacea that produces the quick fix that organisations need. Companies have staff gatherings to motivate. They send managers and supervisors to one day motivational seminars to motivate. But then they bring them back to their workplace where things continue as they have always done. Soon the elation of the "motivational" event fades and they too return to their old habits. More

> **Atmosphere**
> *Leaders create an atmosphere in which others excel.*

often than not motivational events are no more than entertainment. They serve as a quick injection that lifts the atmosphere just as a palliative medicine blocks the symptoms rather than curing the ailment. Pain killers make your back feel better, in the short term, but the pain will return.

What then can leaders do to motivate their staff? The answer is that they cannot motivate them, people motivate themselves. When the atmosphere of the environment in which they are placed coincides with their own desires then self motivation occurs naturally. People are then operating within the higher levels of Maslow's hierarchy. Self actualisation happens when there are no perceived threats and there is enjoyment in what you are doing because it aligns with personal values and objectives. You are very satisfied, self motivated and will excel at what you do. This is not to say that leaders need to do nothing, their job is to create the atmosphere in which others can excel.

Considering again Maslow's model of human needs and his assertion, that the lower needs must be fulfilled before there will be any desire to satisfy the higher needs, I would suggest that the leader's task is to reduce influences on their people that cause them to get trapped in their lower levels of needs. Leaders need to stop de-motivating their followers. Only when this is achieved will they be able to become self motivated. At that point the opportunity to excel becomes a reality and the synergy of effective leadership is realised. Many managers and leaders do not believe that, through their actions (or inaction), they de-motivate their followers. This belief is driven by their unique reality and a lack of empathy; they fail to understand how their actions are perceived by others and how the needs of others may not be the same as their own. They have yet to appreciate the leadership reality. Consider your career so far. Recollect some occasions when you have been de-motivated by your immediate supervisor. Use figure 25 to help identify some occasions that may have applied to you.

> **De-motivating**
> *Never mind motivating, stop de-motivating.*

The list of potential de-motivators is endless and the opportunity to de-motivate is correspondingly large. We all de-motivate others on a regular basis, we tamper, sometimes unknowingly, with the needs of others. The same happens to us and when our needs are threatened, we put up a guard. Our personal defence mechanisms come into play. At these times we feel abused, unwanted or maybe unworthy or ineffective. All these feelings impair performance as our energies become inwardly focused instead of forward focused. We delude ourselves if we believe that we never de-motivate others.

Imagine you have a young officer on your ship or a new supervisor in your department and you are their immediate superior. It is time to conduct the annual or voyage appraisal, you approach the individual and agree a time for the meeting, a few days away. On the agreed day things are madly busy and you decide to postpone the appraisal. You tell the person that you are sorry and that you will get back to them to re-arrange a new time. It remains busy for a few days and you do not reschedule. Eventually you remember the appraisal and slot in the meeting, giving just two hours notice. You are still busy when the meeting time comes but go ahead and get the appraisal finished in fifteen minutes and then rush on to your next task.

Lack of appreciation	No defined career path
Being shouted at	Unrealistic objectives
Receiving negative feedback publicly	Unspecific objectives
Absence of feedback	Blame culture
Being ignored	Unclear instructions
Work conditions	Favouritism
Undue stress	Lack of recognition

Figure 25 Causes of de-motivation

Look at the same scenario from the other person's viewpoint. He is an enthusiastic, committed and ambitious worker and is very pleased when the first meeting is scheduled. He gives the matter some thought and plans to ask you about his future career prospects and for some particular feedback on a project he completed a few months ago. When you postpone the meeting he feels a bit let down but accepts that you are a busy person and that the meeting will be rescheduled very quickly. As the days pass he begins to get concerned and starts feeling disappointed that he is not considered important enough for you to get the appraisal meeting rearranged.

When the new meeting is sprung on him at short notice he is excited but at the same time feels a little cheated that it is being 'squeezed'. The meeting starts and he notes the manager's urgency to get it over with. It feels like the manager is just going through a 'tick the box' exercise. After fifteen minutes it is all over and he has not had any opportunity to ask the manager about his career prospects or to get specific feedback on the project. He leaves the meeting deflated, annoyed and thinking how unimportant his personal concerns, needs and aspirations were to his supervisor. His performance is threatened and his self esteem is reduced. He has been pushed down Maslow's hierarchy, he feels de-motivated and his performance potential has been reduced.

Busy people often forget to consider the significance of their actions and behaviour on others. The illustration above is an example of how even well meaning people can fail to give proper attention to their people. The results can be disastrous.

Leaders need empathy for others, this is not the same as sympathy. Being sympathetic is when you feel that you know how the other person is feeling because you know how you would feel if you were in their place. You have taken your unique reality and assumed that their unique reality is similar. This is a crucial error; we are all different and part of a leader's job is to recognise differences and begin to see things from the real perspective of others. Empathy involves attempting to understand the unique reality of the other person. This allows you to have a better understanding of your impact on them and their perception of the way in which you have treated them. In our example the manager was not necessarily a "bad guy", he was a typically busy individual who did not appreciate the significance of the appraisal to the staff member. He felt, perhaps subconsciously, that if he was the staff member he would have understood that the manager did not have the time to spend with him and accepted that this did not mean the manager lacked interest or concern for him and his future. Small things to managers and leaders are often big things to followers. Leaders who ignore such big things do not create an atmosphere in which others excel, instead they de-motivate and de-tune their staff.

Enthusiasm and passion

Attitudes and moods are contagious and therefore it should be possible to use positive attitudes and moods to develop a motivating atmosphere for others. As leaders we are carefully observed by our followers, they will even copy us on occasions. If we are always late to work then they are likely to adopt this as their approach.

It is important to be passionate about your dreams and aspirations, to be enthusiastic and to set the example for others to follow. Not only will this serve as fuel for your future development it will also provide a continuous catalyst for those around you.

Work with enthusiasm and the habit will quickly be caught by others. This approach is so obvious that you may think it hardly needs to be raised. Look at many of the organisations that you know. Consider the best and the worst. There is nothing more de-motivating to others than a leader who is constantly full of gloom or seems preoccupied with the day to day minutiae. He de-powers those around him. He puts the breaks on the organisation. It then falls behind the competition and as he observes the developing situation he becomes more concerned with the minutiae and his gloom increases. He has created a spiral to disaster.

As leaders you have to be constantly aware that people are watching you, they are observing what you say, what you do and how you do it. You know that you need to focus on getting the most out of everyone and you can only achieve this if the mood you portray becomes positively contagious. It is human nature to want to have a leader with a positive attitude. People feel encouraged when they are well led and their productivity soars, they often try to emulate the leader's style and suddenly an enormous synergy has been achieved. At last the real power of the people is being harnessed.

Enthusiasm comes from the confidence that you can achieve great things.

"Whether you think you can or can not do a thing you are probably right"

In other words, convince yourself that you are going to fail and you probably will. Convince your self that you are going to succeed and you probably will. The latter leads to the enthusiasm of success. Do not forget that whether you are a manager, a ship's officer or a supervisor, the people that work with you are your greatest resource. The future of the unit is in the combined hands of you and your team. The team reflects the mood you set. For the best long-term outcome you need to be respected and trusted and your moods need to be positively infectious. If you can achieve this you will find that everyone shares your passion and with that energised power the vision comes into sight. You are creating the atmosphere in which others can excel.

> **The contagion of moods**
> Moods are contagious whether they are good or bad. Keep your bad moods to yourself, share your good moods.

Optimism

Through the theory of the contagion of moods, optimism becomes an essential element for success and motivation. Any feeling of impending doom must never be openly shown to those around you. Indeed this is a feeling that you must even deny yourself. As a leader you have to inspire and that needs a positive 'can do' approach to every challenge faced. The continual pessimists within any business act as dangerous breaks to advancement and they need to be coached into changing their outlook. If change is impossible then leaders may have to make tough calls. They either have to be persuaded to change or they risk being removed. This is not an area in which you can be too tolerant. The contagion of moods tells us that their constant negative assertions can seriously damage the health of a business that otherwise may have a bright future.

Optimists are always enthusiastic, they believe in challenging objectives, they are "can do" people. As discussed earlier moods are contagious, we catch the moods of those around us and realistic optimists are good to have around. Leaders generate optimism through their energy, commitment and determination. On the other hand pessimism is negative and acts as a break on advancement. It consumes energy, it becomes a barrier, it becomes an excuse for failure, it creates failure. Pessimism has no place in a leader's skills inventory.

I know of a company in India that was facing a huge challenge with its new budget. The Managing Director called one of his senior managers into his office to tell him of his new targets, set by the Global HQ. The manager was shocked by the new numbers since he was very close to his market and knew, within reason, what could or could not be achieved. He complained bitterly to the MD that he had been set an impossible objective. The MD was a practical man and a seasoned leader but he also knew that he was not about to get the HO to ease back on the target. Thinking on his feet he decided to share his real feelings with the manager and said " I hear your assessment, it's a huge challenge but it is the challenge we have to manage. Feel defeated and negative in this office if you must, but the moment you go out through that door you will smile and be positive to all your staff and drive your team towards achieving the objective. And by the way, never let me hear you complain about the budget again, we are going to deliver and I will do all I can to help." The manager left the office feeling that his boss was being both supportive and pragmatic and vowing to use the stretched budget as a positive goal to aim towards.

> **Confidence**
> "A good leader inspires others with confidence in him;
> a great leader inspires them with confidence in themselves" Anon

Filling leadership voids

The leadership role differs depending on where you sit within the organisational hierarchy. A junior officer uses his leadership skills to get the crew to perform cargo duties or anchor positioning in the most effective manner whereas the managing director may use his leadership skills to develop the strategy, direction and culture of the business as it adapts to the changing environment.

A void in leadership anywhere in the management chain will limit the ability of the business to move forward and it will create a wave of de-motivation that ripples throughout the organisation. Such voids exist when any incumbent fails to demonstrate effective leadership. These voids de-power leadership endeavours elsewhere as well as failing to inspire potential followers. The voids make people question the vision and the values of the organisation. All this, at best, does no more than facilitating mediocre performance. At worst it leads to cultural collapse and failure.

> **The leadership void**
> If it is beneath you in the hierarchy start helping it to change. Focus on ensuring that your leadership and that of those beneath you is of the highest standard; then fill any voids.

Leadership voids must be quickly identified, and plugged. Often they can be repaired by effective training or development. This takes time but it is an opportunity that everyone should be given. Alternatively the person may really be in the wrong job and by switching to another position it may be possible to effectively utilise his abilities. Sometimes the development process will not work or it may be impossible to reassign the person. In such circumstances leaders are forced to make tough decisions. Leadership is not an easy task. Occasionally the best leadership decision revolves around the fact that sometimes it is necessary to sacrifice a minority in order to protect the majority. However releasing people from their employment is a final move and should be avoided unless it is the only realistic option. If leadership voids are not identified and filled early, performance will decline.

Once there was a manager working for me running our business in a small but growing country. He was a great person, he would help anyone and put in more hours than you would have thought possible. He was the ultimate loyal and conscientious employee. But his business was going nowhere despite a growing economy and its associated opportunities. His problem was that he was a compulsive administrator. All his time and energy went into the construction and maintenance of a bureaucracy which included the best filing system you have ever seen. Nothing in his business moved unless it was documented, reviewed and approved several times. This was the way he was, he could not be changed. He was in fact a 'control person' and as we have seen, over control is a recipe for business stagnation. Eventually he was transferred to a senior coordinating function with no staff responsibilities where he was able to excel. The leadership void was filled.

Business is getting more competitive, the world is becoming smaller, buyers of services and products have more choice and they are able to impose more demands on suppliers. Look at the decline of the British merchant navy over the last three decades; competitive pressures have led to a wholesale move to flags where manning costs can be kept lower. Suppliers that want to succeed must change (and you are a supplier whether you are navigating a ship through the English channel for your charterer or other cargo interests or taking customers queries in a call centre). You have to accept that increasing competition means sellers, of services or products, have to improve or prepare for failure. If you think your business is in peak condition and that you have nothing to gain by further improvement you are wrong and time will demonstrate this point to you in an unpleasant way.

By working towards the goal of ensuring that every manager or supervisor becomes a leader in his own territory and that there are no leadership voids you will go a long way towards ensuring success while at the same time facilitating positive ongoing change.

De-layering and bureaucracy

These are important atmosphere influencers. People, at the receiving end of excessive levels of bureaucracy usually dislike it. It is incumbent upon leaders at all levels to ensure that bureaucracy is kept to a minimum conducive with the maintenance of essential controls. Unnecessary layers of controls or referrals need to be avoided, they clutter the working world and create noise that detracts from focusing on real objectives. Leaders allow people the freedom to act on their own initiative within clearly defined boundaries linked by vision, values and mutual trust.

Flat organisations work and they work quickly but implicit in them is the need to have empowered performers in every position. They need the delegated authority to act swiftly and the confidence to take appropriate risks. Steep hierarchical pyramids in organisation charts leave too many places for non-performers to hide. Identify the leadership potential in your team. Expose it, empower it, support it and develop it. The objective is to be leaders of business, not kings of bureaucracy. Bureaucracy breeds mediocrity and eventual failure.

An essential element in strong teams is the absence of "bums on seats." These are those staff members who fail to contribute to the progress of the journey and are doing little else but filling otherwise empty seats. In a competitive, well led unit there is no room for passengers, everyone needs to be engaged and contributing towards the vision and objectives.

Passengers exist in many workplaces. They may be resistors who have disengaged following some particular change event. They may be poorly recruited individuals who never really joined the team. The passengers have to be identified, encouraged to rejoin the team or be moved on or out. A passenger in this context damages the whole team rather like a rotten apple in a fruit bowl. Sometimes colleagues are charitable and carry them so they do not get into trouble. This is a false piece of charity as the resultant underperforming team will weaken its own competitive advantage and this may result in the security of more than just the passenger being put at risk.

Poor recruitment produces "bums on seats" but little else. Managers and supervisors at all levels often view recruitment as the realms of the human resources (HR) department. The truth however is that HR are facilitators that assist line management in the recruitment process. Leaders recognise this and consider the recruitment process as one of their major responsibilities.

Failure to recruit only the best will result in excessive wasted cost later. The recruitment process coupled to a meaningful and ongoing appraisal system allows leaders to keep on thickening the cream by continuously improving the quality of the pool of people within the organisation or department. Recruitment then becomes an extremely powerful tool that can dramatically improve performance and add extra octane points to your fuel. The "bums on seats" approach quickly erodes competitive advantage.

> ### Improving the octane number
> *Pay attention to recruitment at all levels and continuously assess and coach existing staff. Don't recruit 'partial fits'. Demand high performance from all staff and accept nothing less.*

Douglas McGregor tells us that people generally want to work and do an acceptable job and yet he suggests that it is us, the leaders, who end up de-motivating them and destroying their desire to work. It is within our power to stop de-motivating others through greater attention to their needs. Leaders that are blind to the needs of others quickly fail.

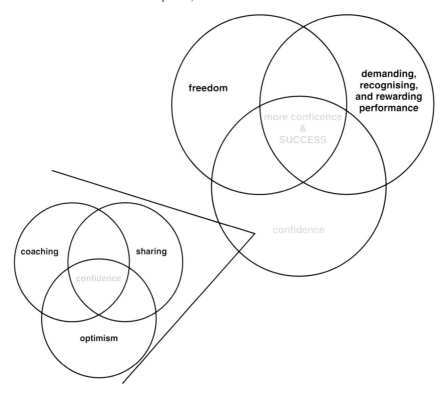

Figure 26 Turn confidence into success

Another important aspect of the exceptional leader is the ability to, sometimes, stand back and accept that others have good ideas, greater understanding or better judgment than they do, in particular situations. They let the voices be heard and they use the wisdom of others. Figure 26 illustrates how leaders give others confidence and that this coupled to allowing people the freedom to lead can bring collective success in a motivating atmosphere.

Chapter 12

Action: Developing a leadership culture

Summary

This chapter introduces the final finger of the hand of leadership. Here we look at what leaders really can do, rather than what they say they can do. Coaching, sharing and managing change are discussed as areas in which leaders need to be active. The need for leaders to demand and acknowledge success is advocated as is the need to constantly challenge others to achieve more. These assertive traits are set against the need to give others the freedom to lead and to let them take the credit for their successes, while the leader always does what he says he will do. It concludes by discussing the leader's responsibility to take tough action whenever it is necessary.

The Hand of Leadership

- **Awareness** *of self / of others* → Tuning behaviour and using empathy
- **Direction** *defining the route* → Setting objectives for self and others
- **Openness** *sharing the messages* → Enabling two way communications
- **Atmosphere** *reducing de-motivators* → Creating the mood for self motivation
- **Doing** *Self / others* → Taking action

Survival versus excellence

It seems obvious to suggest that a leader should aim to get the best out of his people with a view to achieving the longer term objective. And yet many leaders seem to miss this simple concept. In reality the only times when this is not an essential requirement for success is when an emergency situation is being faced and survival is all that matters. Excellence, part of the vision of any sustainable leadership initiative, is not the same as survival. A business in distress usually requires more aggressive leadership and this seldom permits the building of a sustainable leadership culture. Having recovered the distressed business the leadership can turn its focus towards longer term sustainability and this is when a lasting leadership culture can be developed, only then will it be possible for real and sustainable excellence to raise its head.

Excellence is only possible when people are active. Words alone will never create excellence, just as leadership will never exist without all the fingers of the Hand of Leadership being present.

Sustaining leadership

Sustainable leadership does not give everyone the opportunity to relax and become complacent. On the contrary, it is a requirement for leaders to constantly challenge their people, always pushing and guiding them to operate beyond their normal expectations. Leadership relies on the realisation of constant, competition beating change. Leaders need energy and the ability to drive others to success. Taking the message from page 119 that moods are contagious it is possible to understand why a leader's display of energy is important in any endeavour to develop exceptional levels of performance in others. One of the key attributes of leaders is that they are energised with a desire to succeed and they realise that to do this they have to bring the majority of people along with them.

Leaders help others to excel through their interactions with them and through the atmosphere that they create through their operational style. Some of these interactions that facilitate sustainable leadership are discussed below.

Coaching

Leaders have a number of skills and an amount of knowledge that they can share with others through coaching and mentoring. They share their expertise and experiences and encourage others to develop by building trust and confidence. Coaching also strengthens the leaders own self awareness as they explore how they can use their experience and behaviour to help others develop. The mode of coaching can be direct and 'on the job' or indirect through more general encouragement and support. As part of the coaching process leaders demonstrate that they care and are prepared to allow and help others to develop. These leaders are masters at the art of situational leadership (see page 59). They are able to adjust their style of interaction during the process, as the situation and context changes, in order to assist the coached person to maximise their learning.

To lead in a sustainable fashion you must 'care' for the people that you work with. If you do not have that degree of compassion you cannot become a sustainable leader. Be careful not to confuse being caring with being soft, the best leaders care for their people while at the same time having the ability to be tough when it is called for. They always remember that they are not better people than their followers, they are just fulfilling a different role.

The process of development leads to mistakes and this is one of the most valuable learning tools available to us all. A good coach gives people the confidence to experiment. At the same time he makes it clear that as long as they perform within the boundaries established by the values of the organisation and that they are moving towards the vision or a particular objective they will support them in success and failure. Let them make some mistakes and expect them to learn from them.

Leaders understand that in order to move forward risks have to be taken. Yet

> **The measure of success**
> "...is not whether you have a tough problem to deal with, but whether it's the same problem you had last year."
> John Foster Dulles

through coaching and mentoring they manage the level of risk involved. As the eureka point (see figure 22, page 104) is reached the people involved suddenly see that they are starting to operate at a level beyond their previous expectations.

Sharing

Part of the coaching process is the sharing of knowledge, information and experience with others. Some people find this very difficult. If you are one of those people you need to reflect on why and on what you are going to do about it. Leaders do not fear sharing, on the contrary, they encourage it. Usually a reluctance to share is a sign of insecurity, a sign of feeling threatened. Yet "none of us is as good as all of us". If a leader tries to do everything himself he will eventually fail, leadership depends on willing followers, not robots. A reluctance to share because you imagine you will become threatened or insecure will ultimately turn the imagined into the reality. Sharing knowledge and experience is a win/win activity that can bring huge personal and professional satisfaction. Failure to share leads to failure, for you and your team.

Handling change

We live our lives in constant change. We change as we grow, we change as we acquire new knowledge and skills, we change as technology advances. How we adapt to change and its opportunities determines our success. Imagine if you had not acquired your mother tongue during your formative years, you would be at a serious disadvantage compared with those that had. Alternatively consider how you would have been able to cope today if you had refused to learn to use a computer. Leaders are masters of recognising the early need for change and managing its implementation, they are agents of change. Learning the skills and having the willingness to recognize opportunities for change is essential for every leader.

Implementing change, such as a new strategy or process, is one of the hardest tasks for any leader to action, it is also one of the most common. The reality of change is captured in the two quotes below.

"The only certain thing in business today is change"
anon

*"There are many ways of going forward but
only one way of standing still"*
Franklin D Roosevelt

Change is what business leadership and industry leading success is all about at all levels. No one can avoid the necessity to constantly change. Successful leaders implement change fast and early; they use change to enhance their competitive advantage. If you think that this is just a novelty that will disappear with time you are wrong. Embrace change or you will lose the ability to compete and you will soon become obsolete.

If you and your team can do something new or better that works consider how long it will be before your competitors mirror you? Keep a constant rhythm of change; make it a way of life for everyone that works with you. Do not forget your aim is to ensure that your competitors are copiers not innovators and that is where you gain real advantage, speed and dominance. It is true that some change initiatives will fail but to do nothing always leads to failure. To do something that fails leads to a learning experience and to do something that succeeds fuels greater future success.

> **Make your competitors copiers**
> *Drive change hard and fast. Continuously innovate and search for new ways of doing things. Then you will stay ahead.*

People tend to defend their comfort zones, they do not always like the uncomfortable feeling that change can create. This natural reaction presents leaders with a challenge, how to progress change without disengaging their followers? In recognition that change leads to discomfort leaders need to direct their efforts in three ways:

(i) Establishing and communicating the objective of the change, in a way that is clear and compelling. Given a sensible rationale for the change people are more likely to accept the need for the journey

(ii) Supporting people through the change process with a level of guidance that is appropriate to the individual. Everyone needs to be treated differently if they are to end up feeling the same. Passing through the pain of change takes time and people need to be assertively encouraged rather than aggressively forced through the process

(iii) Monitoring, listening and measuring progress. Always with the willingness to change direction if it has the potential of delivering a better outcome. Change changes.

It should be an obligation for all managers and supervisors to regularly move outside of their comfort zone and to ensure that their followers share in the same experience.

The whole purpose of change is to maximise opportunities. Change gives us a new opportunity every second of our lives. Embracing change as a high-octane fuel will serve to accelerate you beyond your competitors.

> **Fearing change**
> Fear change in business and you will fail. Change is a way of life for tomorrow's leaders.

Resistors to change exist and leaders need to be aware of their various incarnations. All resistors disengage from the objective of the change. We will call them "quitters" and they can be grouped under three headings:

1. Those who disengage but subsequently re-engage – they quit and rejoin
2. Those who disengage and leave – they quit and resign
3. Those who disengage and never re-engage - they quit and stay

The first type is the person who is disturbed by the notion of change and initially resists. Given time and effective leadership from his supervisor he returns to the fold and embraces the new paradigm. This is the quitter who rejoins. The second type is the person who cannot and will not handle the change no matter how supportive or compelling his leaders are. Eventually he resigns or has to be reassigned. This is the quitter who quits and leaves. The final resistor is the most dangerous, this is the quitter who stays. He rejects the rationale for change and works overtly or covertly to derail the process. This type of resistor needs to be identified, encouraged to rejoin or removed. Leaders have to make some tough calls since one of their responsibilities is to protect the employment of the majority, sometimes that means parting company with the minority.

Demanding and recognising success

As managers and supervisors leading groups of people you have a right and duty to demand results from all who work for you. You also have a duty to recognise and reward success.

A key to the tasks associated with demanding success is the clarity given to the objectives. If people do not know where they are aiming they are unlikely to hit a bull's eye. Subjective targets should be avoided since they are open to interpretation and debate which can result in time wasting and the generation of unnecessary ill feeling. Measurement of success against objectives should be unambiguous.

Objectives need to be challenging and always discussed with staff. Their commitment to them must be secured or your task remains incomplete. There is no room for challenging objectives to be accepted in a cynical fashion "ok we'll say yes, but he must be mad, we'll never get there." It is always surprising to see how good people rise to a challenge and often turn unbelievably difficult objectives into reality. Once they are convinced that with extra drive, a little luck and real effort challenging smart objectives can be met. The level of initial discomfort that they experience will quickly decline. Your job then is to bring your staff to this level of awareness. When commitment has been secured you need to work with them to formulate a plan to drive the business towards the objective.

The issue here is that properly managed challenging objectives serve to focus and motivate people to take themselves beyond their comfort zones and achieve exceptional performance. Mediocre objectives can only at best produce mediocrity.

A few years ago I attended a formal series of presentations by a number of my direct reports (line managers) in the Far East. They were making their presentations to the new group CEO who was visiting from Europe. One unit manager proposed a stretch target for the coming year showing about 30% growth. He went on to discuss in a positive fashion his local market place and how he should be able to achieve the growth. The perceptive CEO saw that even 30% growth represented a soft target in the prevailing market. He interrupted the presentation with a request that the manager accepts a 'stretch dream' target representing 100% growth. The manager turned very pale. The CEO pressed the issue and the manager, hardly surprisingly, conceded. But the CEO had not finished. He said 'now I want an email from you today, no more than one page long, detailing what you need to make this new revenue target'. The manager produced the email.

As his local manger I agreed to his capital requests. The line manager had a restless weekend but by Monday he was overflowing with enthusiasm and sharing his new target with his staff. Two months later the budget (a top down process in this particular organisation) arrived and the line manager saw that his budget objective was appreciably lower than the 'stretch dream' set with the CEO. By this time however revenue and profit was growing well and the line manager was feeling delighted that he was clearly going to be able to exceed his budget and get quite close to the 'stretch dream'. He and his team were performing beyond expectations.

At least three leaders came into play here:
- The CEO – created the vision and challenged the line manager to deliver

- The Line Manager – accepted the challenge
- The Local Manager – gave the line manager the freedom to act

Once success has been achieved it needs to be recognised. Too often managers and supervisors focus their energies on recognising only poor performance. Leaders address poor performance but they also attach a lot of importance to recognising and rewarding success. Ken Blanchard, a co author of *The One Minute Manager*, advocates the importance of managing by walk about. By this he means that mangers and leaders should be accessible and regularly engaged in the front line of whatever their business is. He goes on to say that time is best spent looking for people doing the right thing rather than the traditional management approach of looking for people doing the wrong thing. Whenever you see someone doing the right thing, offer them some praise and encouragement, let them know that you appreciate that they are doing a good job. This is recognition and for most it is also a reward. Money is not the only way of rewarding a job well done.

Challenging others

Challenging others is one of the key drivers used by leaders to perpetuate change and improvement. Since change is one of the few constant components of tomorrow's successful business, leaders recognise its significance and power. Leaders forever challenge the status quo. They are seldom happy to accept life in the comfort zone for themselves or for anyone they are involved with.

I know one leader who loves expensive sports cars but he insists that he needs a car that will sometimes breakdown. He wants the sporting thrill but he also wants the uncertainty and challenge of something that is operating on the edge, outside of the comfort zone. He accepts the possibility of failure but finds the probability of success much more thrilling.

One of today's leading management authors Stephen Covey says 'If we keep doing what we're doing, we're going to keep getting what we've always got.' I believe the prognosis is even worse, 'if we keep doing what we have always done it will not be long before we have nothing to do.' To stand still is to go backwards and to go backwards is to be on the path to failure.

> *Failure*
> *Fail occasionally and your leader should coach and encourage you. Fail regularly and your days are numbered!*

Challenging others must be a daily feature of your approach to your work and your work colleagues, nothing should be accepted as perfect. At first people will be uncomfortable with this constant 'what if' pressure. But before long, if you have established mutual respect and trust, they will begin to think outside of the box and new opportunities will be opened. They will become innovative, more entrepreneurial and start challenging their own staff; a real case of synergy in action. Recognise their successes.

The whole challenge issue is inextricably linked to change. First you challenge, then you change. Leaders have a need to be (sensibly) ruthless. Some do not have the energy and drive to follow through this challenging issue. This absence of drive and the unwillingness to challenge can arise for a number of reasons, perhaps an over-riding desire to be liked or a social conscience that tells them not to take tough actions that will result in job losses or a relaxed " let's not rock the boat" approach.

> **Determination**
> The most influential and powerful individuals are not necessarily the most talented, but they are the most determined

Letting others take the credit

The best leaders let the results speak for themselves. They don't seek praise. They know when they have done a good job. Leaders that steal the praise for success from others are unlikely to continue receiving similar levels of support in the future. They have abused the trust of their followers because success is only possible with team effort. People are able to move up Maslow's hierarchy when they feel valued and appreciated. Be sincere with praise and seize every possible opportunity to show that you value your people.

Walking the talk

No leader can be successful unless he includes in his armoury of skills those of driving action and completion. Everyone must walk the talk, always. It is no use simply postulating an approach; there must be follow up, there must be action, the words must be implemented. As you go about the task of leading others and allowing them to become leaders, it is essential that no stone is left unturned and no promise is left unfulfilled. Leaders are doers, not just talkers. Talkers who fail to do what they say they are going to do should be high on any leaders list of undesirables. Empty promises are cheap, leaders and committed followers get things done.

This is an obvious fact, yet it remains unimplemented by many. How often have you encountered people that find it is easy to say what needs to be done but seldom do anything else about it? These are not just "the professional committee types", they are often line managers and supervisors. They are people who think words speak louder than actions, they are often the loud voices that Plato so despised and Hannah Arendt observed.

Delivery is not optional, it is a way of life. Delivery is what we are all paid for. Leaders cannot afford to tolerate repeated non-delivery. Dealing with underperformance is one of the challenges they face. The better the leadership, the less often this challenge will arise.

The freedom to lead

Rules, policies, approvals, controls are all necessary but they can become overwhelming. Unnecessary bureaucracy and control stifles growth, consumes energy and endangers morale. It is impossible to lead by doing everything yourself. Leaders must delegate and empower others to act, free from unproductive hindrances.

> **The orchestra**
> The conductor delegates the playing of the instruments to others, but he leads them all to success

The people that sometimes find it difficult to share do not allow others to have the freedom to make their own decisions. Giving this freedom builds confidence and confident people who respect their leadership are usually the ones that produce exceptional performance.

Tough action

As a leader you have a number of key challenges including delivering value to the owners, taking the team towards its common vision and protecting the employment of the majority. Another way of expressing this last responsibility is 'being prepared to sacrifice the employment of the minority'. The best organisations and the best leaders know that the issue of under-performers needs to be addressed. They realise that they cannot build and sustain a 'best in industry' operation unless you have best in industry people. And since the aim is to push through to the eureka zone as quickly as possible there is a need to act quickly when you have the wrong people. In the end this will be to the benefit of the majority.

There can be no escaping from the fact that in developing your leadership skills you have to be able to become sensibly ruthless when the circumstances dictate. You have to learn to make tough calls when they are necessary. Being sensibly ruthless is the ability and determination to look at every set of circumstances in a dispassionate manner and take the action necessary to ensure that the organisation is not either derailed or unnecessarily delayed in its journey towards it vision.

Power ≠ Leadership

Too often people in leadership positions consider themselves to be absolute monarchs. Leadership in any workplace is not a birthright. The motives of such leaders are usually far from altruistic. They are more likely to be based on egotistical and selfish needs. Perhaps through the ability to think on their feet, to portray an aura of supreme confidence or the good fortune of being born with a charismatic presence many mistake them too easily as superior leaders. The test of their entitlement to the title of leader is whether followers willingly follow them over time or whether they are followed because they wield some form of power over their people.

Beware of leaders who use threats as their main management tool. The followers are unlikely to be either enthusiastic about their work or willing to produce exceptional performance. In the long term leading by threatening behaviour is not a sustainable route. People become disengaged and their focus turns to survival. They rapidly descend Maslow's hierarchy and performance becomes a casualty. When staff do not have a high level of confidence in where they are being taken (vision/objective) and the rules of how to get there (the values) they will collectively under-perform.

> **People decisions**
> "Of all the decisions an executive makes, none are as important as the decisions about people, because they determine the performance capacity of the organisation"
> Peter Drucker

A recent survey has found that few UK companies are leveraging the goodwill that can be offered by their staff. The MCA Communicates (now part of the WPP Group) survey, *The Buy-in Benchmark 2002*, reveals that only just over half of those working in larger UK companies are currently committed to helping their organisations succeed. Whereas the level of performance in such organisations is open to debate there is no doubt that the lack of commitment and goodwill restricts the potential that could be realised. The people have the power and they are the key resource that effective leadership harnesses.

Conclusion:
Embracing change

Summary

This chapter discusses the rate of change in today's business world. It reflects on the consequences for organisations that fail to embrace the opportunity which it presents. It continues with an assertion that leadership is becoming an important tool in the race to improve competitive advantage between organisations and business units of every kind. Leadership is a function of the majority rather than a privilege of the minority. It reiterates that leadership at every level is an acquired skill and that it is within the reach of all. Leading edge ships, fleets and business organisations need future staff that see leadership as an integral function of their position. It concludes by repeating the proposition that Leadership Throughout is the way forward, since the people really do have the power.

Change

Leadership is about change. It is a process that directs people and organisations through continuous evolution. Leadership should be constantly present. It should not be an optional extra that may be considered useful on one day yet unnecessary on the next. It is a process that renews and adapts itself as circumstances alter.

Leadership development must become an important component of any manager's ongoing learning process. If this is to be operationalised it has to be part of an organisation's cultural identity. The development and use of leadership skills throughout the organisation must be supported as an ongoing practice,

The merchant service, like any area of commerce and industry, faces a rapidly changing environment. Darwin's notion of evolution has accelerated with the coming of the global village. The consequence is that today we are seeing evolution occurring at the speed of revolution. Traumatic though this may be for some it is a fact of life.

Consider these facts:

Twenty years ago the internet was little more than a database for intellectuals, today it is the most prolific information medium in the world.

Twenty years ago we had not heard of Aids, today it is the scourge of many societies.

Twenty years ago mobile phones were just beginning to appear in the developed world albeit that, along with their batteries, they weighed several kilograms. Today if we forget our mobile we are left feeling vulnerable and exposed.

Twenty years ago the Berlin Wall separated east from west. Today it is gone and the Chancellor of a unified Germany comes from the east.

Twenty years ago Dubai was a modest city on the shores of the Arabian Gulf, today it is said to be the temporary home of half of the world's construction industry cranes.

Who in 1987 would have predicted these outcomes? Why should we then suppose that there is no need to change and that we can just continue as we are without any threat? The reality, in fact the leadership reality, is that we must change or we will not survive. The greater the meaningful change the better our quality of survival will be. A competitor who changes faster than you threatens your very existence. Simply to change is not enough, you have to change faster than the rest. This is the challenge of effective sustainable leadership. Today leadership needs to be widespread throughout organisations. In the past success may have been achieved by weight of numbers, today it comes through the collective creative energy of groups of people that are led so that they can excel.

Comfort zones need to be the base where energies are recharged, they need to be visited regularly but they should never be your permanent residence. Through personal growth and increased self awareness it is possible to achieve a higher degree of freedom of choice. Leaders create their own success and in doing so they create success for those around them. There is no free ride. Discomfort is a component of change. However the discomfort of failure is far more painful. The more successful you are the more life choices you have and the more successful those around you will become.

Most people need to redirect themselves from time to time whether within an organisation, perhaps by taking on a new and more challenging role, or by making a complete redirection in their lifestyle. What is abundantly clear is that you have to get on with life to make it as fulfilling as possible. You have to make things happen and be the author of your own masterpiece, whatever that may be.

> **The comfort zone is dangerous**
> Your comfort zone will not challenge you, remain in it for too long and your value to yourself and others will decrease.

Psychologists take several different views of why we behave and develop the way we do and there has been much debate about how much choice we actually have and whether our 'self' is publicly or privately shaped. In adult life, it seems clear that it is predominantly from our social world that we gain our ideas, understandings, hopes and aspirations. Human beings are great copiers and through this approach we observe and develop an understanding of others and the things they achieve. Clever people are selective in who and what they copy. They choose to acquire skills from the best. Leaders are not born, they are developed and they develop themselves through observation, experiment, action and constant learning through whatever opportunities and mediums they can access.

Our actions determine what will happen. Whereas we cannot ignore the influence of society and culture on our freedom to choose it is nevertheless true that we each have a huge ability to shape our own future. There will always be limitations but if we fail to develop we fail to change and if we fail to change we will be failures. Leaders are successes not failures.

Creating change is an interactive process that encompasses all the fingers of the hand of leadership, awareness, direction, openness, atmosphere and doing. Together they become leadership, individually they are not.

Is there risk in the future?

With evolution occurring at the speed of revolution there is bound to be risk. Expectations for each of us are greater than they used to be. For example, companies and governments are no longer willing to be absolutely loyal to their staff. Employees must prove their worth in the market place. Perhaps you can recall the abolition of the Dock Labour Scheme in the late 1980s when Margaret Thatcher's government ended the "jobs for life" situation that had become a given in the ports of the UK after the end of World War Two. The employer has a major responsibility towards all their stakeholders, not just to their employees. If you cannot help to fulfil the expectations of the successful companies of the future there will be no place for you in such organisations. The future is going to be an exciting place but it is also going to be a terrifying and unpleasant place for those not prepared to embrace change. The vanguard of seafarers and other employees, at all levels, who take up the challenge of change have a huge opportunity for success. Ignore change at your peril.

Leadership is no longer the realm of the CEO or Captain alone, it has become a function of the majority. Supervisors, junior, middle and senior management all have a responsibility to display effective leadership to those around them. Future success will come to those who are able to leverage the skills of the majority through dialogue and example rather than through command and control. Leadership at every level is an acquired skill which is within the reach of all.

> ### Be daring!
> *'If you do things well, do them better, be daring, be first, be different'*
> Anita Roddick, founder of Body Shop

Companies, professional associations and government organisations are all bound in a web of interacting influences. They need leaders to unravel the web and develop, coordinate and implement effective outcomes. Merchant fleets that excel in the future will be manned by leaders at all levels who have strong technical and strong leadership skills. They will be people who embrace change and believe sincerely that their future success lies in the hands of those that follow them.

People are the key to creating successful enterprise. People really do have the power.

Bibliography

COLLINS, J. 2001. *Good to Great.* New York , USA: Harper and Collins

GOLEMAN, D. 2006. *Social Intellegence – The New Science of Human Resources.* New York, USA: Bantam Books.

GRINT, K. ed. 1997. *Leadership. Classical, Contemporary and Critical Approaches.* Oxford, UK: Oxford University Press.

HERSEY, P., BLANCHARD, K., JOHNSON, D. 2000, 8th Edition. *Management and Organizational Behaviour: Leading Human Resources.* New York, USA: Prentice Hall.

HERZBERG, F., MAUSNER, B., SNYDERMAN, B. 2005 (eighth printing). *The Motivation to Work.* New Brunswick, USA: Transaction Publishers.

KANTER, R., M. May 2000. Ivey Business Journal. *The Enduring Skills of Change Leaders.*

MACHIAVELLI, N. 2005. Bondanella, P., Ed. *The Prince.* New York, USA: Oxford University Press.

McGREGOR, D. 2006 (1st published 1985) *The Human Side of Enterprise.* McGraw Hill

McGREGOR, D. 1967 *The Professional Manager.* New York, USA: McGraw Hill

MINTZBERG, H. (1977) *The manager's job: folklore and fact.* Harvard Business Review, vol.55, July-August.

MORRELL, M., CAPPARELL, S. 2001. *Shackleton's Way.* London, UK: Nicholas Brealey Publishing.

SENGE, P., KLEINER, A., ROBERTS, C., ROSS, R., SMITH, BRYAN., 1994. *The Fifth Discipline Fieldbook.* London, UK: Nicholas Brealey Publishing.

SENGE, P., SCHARMER, C., JAWORSKI, J., FLOWERS, B.. 2004. *Presence; An Exploration of Profound Change in People, Organisations and Society.* New York: Doubleday

McLUHAN, M., 2001 (originally published 1964). *Understanding Media.* Oxford UK: Taylor Francis Group Ltd.

KOTTER, J., 1996. *Leading Change.* Boston USA: Harvard Business School Press

Appendix 1

Communicating with clarity

1. The garage owner was a man — **don't know**
2. The compressor was electrically driven — **don't know**
3. Someone turned on the electricity — **yes**
4. The thief stole the car and drove it out of the garage, just missing the door — **don't know**
5. The car nearly hit the garage door because it was being driven fast and the doorway was narrow — **don't know**
6. The police will try to arrest the thief and return the car to the garage owner — **don't know**
7. The garage owner called the police — **yes**

Explanations:

1. Nothing in the story tells you that the garage owner was a man.
2. There is nothing written confirming that the compressor was electrically driven.
3. The owner turned on the electricity. Everyone is "somebody".
4. You do not have enough information to determine if the car was stolen.
5. Nothing in the story confirms that the car is actually in the garage at any point.
6. The police may choose to return the car to its owner rather than the garage owner.
7. The garage owner did call the police.

INDEX

A

action 7, 18
active listening 94
aggression 30
aggressive 73
allowable weaknesses 47
altruistic 134
Arendt H 2
Argyris, C 52
Aristotle 103
aspiring leaders 30
assault 33
assertive 73
assumptions 53
atmosphere 7, 8, 16, 115
attitudes 118
autocratic 13, 14
Awareness 7, 12

B

bad leadership 7
battles 107
beer 58
behaviour 111
behaviour tuning 50
behaviour types 73
behavioural scientists 40
Belbin Dr. M 47
Belbin Team Roles 47
belief system 53
belonging 79
Bennis W 23 27
blame 110
born leader 2
Briggs M 49
bureaucracy 115

C

causing harm 109
change 127
charisma 30
Churchill Sir Winston 59
circle of choice 57
clarity 105
climb to conflict 52
coach 9
coaching 63
Codrington Sir William 3
Collins J 67
comfort zones 136
command and control 26
commitment 60
common understanding 110
communicating 89, 93
communications 7, 80
competence 60
competitive advantage 26
confidence 30
consistency 99
contagion of moods 119
contingency based leadership theories 58
continuous improvement 41
corporate visions 26
Covey S 51
creating Concern 109
cultural 52
cultural diversity 52
culture 51

D

'Daring Dialogue' 93
Darwin C 135
de-layering 122
de-motivated 112
de-motivators 16

debate 107
defining leadership 23
DEFRA 19
Delegating 9, 61
Demanding...129
departmental visions................................26
dialogue 107
dictatorial 13, 14
difference 49
directing...61
Direction...7, 13
discipline..56
distributed leadership..............................64
diversity 28
Dock Labour Scheme 137
Doing 7
duelling 111
dysfunctional organisation 89

E

egocentric behaviour................................30
emotions 111
empathetic...47
empathy 7, 47
empower 9
engagement...98
Enron 29
enthusiasm...118
environment...19
ethics..19
eureka zone 104
excel 115
excellence..8, 9
extrovert 2

F

favouritism 32
feedback...112
fight...93, 109
Flag for Dialogue 108
fleets 27
flight..93, 109
flight or fight 107
Flores F 93
freedom 106

G

George B 37
German leadership 31
global village..52
goal(s)...47
goals 82
Goleman 43
good leadership 7
Gore W L 67
Gosling J 96
great man theory 58
Greenleaf..67
Grint K 2, 76

H

Hand of Leadership...................................7
hard skills...55
Hersey and Blanchard 57
Herzberg F..16
honesty..88
human resources...................................122
humility..40
hygiene factors 16

I

impatience 80
inconsistent..80
indirect leadership 10
integrity..102
ISO 9000 41

J

Jigsaw@work 49
Journal of the Honourable Company of
    Master Mariners 3
Jung C 36

K

Kanter 58
Klein 65
knowledge 29
Knowledge @ Wharton 37
Kotter J 26

L

ladder of inference 51
laissez-faire 23
Lao Tzu 65
leadership 23
leadership circle 68
    is lost 26, 94
    organisation 11
    organisation chart 7
    pyramid 7, 10
    reality 49
    skills 1, 55
    therapy 56
    Throughout 7, 10
    voids 120
Lee Kuan Yew 56
liars 103
listening 88
loud voices 29
loudest voices 2

M

Machiavelli N 66
Marine Insurance Acts 25
mariners 1
mask 46
Maslow A 77
Maslow's Hierarchy of Needs 16

McGregor D 35
Merchant Navy 3
Middle East 28
military conflict 25
Mintzberg H 59
MIT's Sloan School of Management 38
monologues 112
moods 43
motivating factors 17
Motivation to Work 16
motivation 7, 115
motivators 16
multi-cultural leadership skills 52
mutual 111
myth 30

N

Nader R 34
Nautical Institute 9
noise 32
Northouse 67

O

objectives 7, 85
openness 7, 14
operational leadership 10
optimism 119
optimists 120
orange zone 111
organisational development 39

P

passion 118
passive 73
perceptive awareness 50
persona 46
personal agendas 32
    change 80
    development 9
personality 46
Plato 2

politics 32
Pyramid of leadership 10

R
realistic optimism 32
recognition by others................................16
recruitment 122
Rees M 49
respect...102
restructure 33
robots 37
Roosevelt F D 28, 128
ruthless 132

S
safety 79
self awareness 43
self esteem 79
Senge P 52
servant leader 67
Shackleton Sir Earnest................................1
sharing 127
shipmasters 25
ships 27
situational leadership.................................57
situational skills 2
soft skills 1
SMART 85
STCW convention 9
stereotype...12
stereotypical leaders..................................30
strategic leadership 10
Style...80
success recognition of.............................129
supporting 63
sustainable 7
leaders..9
synergy 11, 132

T
Taoism..65

team leadership...10
teams 32
technical skills 8
Thatcher M..137
third truth..55
time management 90
timeliness...88
tough action 133
trait model..58
transactional 66
transformational 66
transparency 105
trust...88, 102
turnaround experts....................................33
tyrants..31

U
Ugandan leadership 31

V
values...88
Videotel..9
visionary leadership 27
visions..83

W
Wolfowitz P 101
Woods T 30
World Bank..101
Worldcom 29

X
X and Y Model 38

Maritime  Futures

Maritime Futures is the concept behind a new series of books which are designed to expand the horizons of the possible, and explain how the work of innovative technologies will impact on maritime industries. The series aims to demonstrate how management strategies can be aligned to take advantage of new opportunities.

Each book makes a ground breaking contribution to the understanding of new technologies and the way they can be applied to improve performance.

:/ Waves of Change — inspiring maritime innovation

This book explains how conflicting expectations arise between those who hold traditional values and those who see innovation as the way forward. Sooner or later, shipping will have to embrace the wider opportunities created by new developments in information technology. The book sets out to demonstrate how managers, and that includes sea staff, can transform this potential into results. Dr John Robinson, himself an experienced senior executive, provides an enlightening account of how to implement innovative strategies A truly scene-setting book for the series that will change perceptions forever.

:/ Safety Management

This book explores the ways that different industries have adapted to ensure their operations are safe. The author Professor Chengi Kuo, a ship technologist, explains the difference between traditional prescriptive regulations and the discipline behind the safety case approach. In an innovative environment the past is not necessarily an infallible guide to the future, but ways have to be found to protect the public, workers and the environment from unsafe practices. Ultimately, whichever system is used has to be managed and the author encourages readers to question whether it is now time to rethink our priorities.

:/ Integrating ship bridge systems — Vol 1 Radar and AIS

Marine equipment has been designed and produced as separate units, but the availability of fast reliable data processing now enables individual items to he integrated into more flexible and applicable work stations. Linking Automated Information Systems to radar (compulsory on all new radars post 2008) provides new opportunities for presenting ship data so that more informed navigational decisions can he made. With integration comes more variations of possibilities. Dr Andy Norris, a leading equipment designer, shows how to avoid the clutter of spurious information by concentrating on the navigational tasks for which the equipment will be used.

:/ Integrating ship bridge systems — Vol 2 Positioning

Although all ships use GPS, there is still no alternative system at sea to take over if the GPS satellites are disabled. For reliability there need to be three systems available. With only two it is not possible to know which system has failed: so the search is on for a three-way electronic solution to this age old maritime problem. Dr Andy Norris explores the viability and benefits to the industry of making that crucial step into the electronic age. At this stage the sextant will surely follow the example of Mr Morse's communication code. Further volumes are being planned.

:/ Leadership Throughout

Who is going to win the tug of war between those who have their feet firmly on the ground supporting traditional maritime values or their more cerebral adversaries who see opportunities for mobilising support through new alliances to capitalise on the opportunities presented by powerful information technology networks.

It is an intriguing question which all managers and sea staff will have to address as change gathers momentum amplified by the economics of competition. Leadership in this context assumes new meaning with respect to organisational development. However, there can be no point in being out in front if everybody else is sailing in a different direction.

The author is a mariner, with an MSc from Exeter University, and is a personel development consultant active in the multi-cultural Middle East. He doesn't say leading is easy but he does demonstrate how to acquire the skills to succeed.

:/ Other subjects being planned

Educational access, the key to success in taking computer based systems to sea — Expanding the economic horizons of sea transport — Redefining Maritime Policy.

Maritime Futures

are published by　　　　　and supported by

The series is supported by the UK Maritime Forum
which includes the following professional associations:

The Royal Institution of Naval Architects
The Royal Institute of Navigation
The Institute of Marine Engineers, Science and Technology
The Institute of Chartered Shipbrokers
The Nautical Institute
The Society for Underwater Technology

If you would like to be kept in touch when new titles appear,
please simply record your e-mail address by sending it to
maritime.futures@nautinst.org

For further information contact
www.nautinst.org
The Nautical Intstiute, 202 Lambeth Road, London SE1 7LQ
Tel: +44 (0)207 928 1351

Notes

Notes

Notes